Nuclear Risk in Central Asia

NATO Science for Peace and Security Series

This Series presents the results of scientific meetings supported under the NATO Programme: Science for Peace and Security (SPS).

The NATO SPS Programme supports meetings in the following Key Priority areas: (1) Defence Against Terrorism; (2) Countering other Threats to Security and (3) NATO, Partner and Mediterranean Dialogue Country Priorities. The types of meeting supported are generally "Advanced Study Institutes" and "Advanced Research Workshops". The NATO SPS Series collects together the results of these meetings. The meetings are co-organized by scientists from NATO countries and scientists from NATO's "Partner" or "Mediterranean Dialogue" countries. The observations and recommendations made at the meetings, as well as the contents of the volumes in the Series, reflect those of participants and contributors only; they should not necessarily be regarded as reflecting NATO views or policy.

Advanced Study Institutes (ASI) are high-level tutorial courses intended to convey the latest developments in a subject to an advanced-level audience

Advanced Research Workshops (ARW) are expert meetings where an intense but informal exchange of views at the frontiers of a subject aims at identifying directions for future action

Following a transformation of the programme in 2006 the Series has been re-named and re-organised. Recent volumes on topics not related to security, which result from meetings supported under the programme earlier, may be found in the NATO Science Series.

The Series is published by IOS Press, Amsterdam, and Springer, Dordrecht, in conjunction with the NATO Public Diplomacy Division.

Sub-Series

A.	Chemistry and Biology	Springer
B.	Physics and Biophysics	Springer
C.	Environmental Security	Springer
D.	Information and Communication Security	IOS Press
E.	Human and Societal Dynamics	IOS Press

http://www.nato.int/science
http://www.springer.com
http://www.iospress.nl

Series C: Environmental Security

Nuclear Risk in Central Asia

edited by

Brit Salbu
Norwegian University of Life Sciences,
Department of Plant and Environmental Sciences,
Ås, Norway

and

Lindis Skipperud
Norwegian University of Life Sciences,
Department of Plant and Environmental Sciences,
Ås, Norway

Published in cooperation with NATO Public Diplomacy Division

Proceedings of the NATO Advanced Research Workshop on
Radiological Risks in Central Asia
Almaty, Kazakhstan
20–22 June 2006

A C.I.P. Catalogue record for this book is available from the Library of Congress.

ISBN 978-1-4020-8316-7 (PB)
ISBN 978-1-4020-8315-0 (HB)
ISBN 978-1-4020-8317-4 (e-book)

Published by Springer,
P.O. Box 17, 3300 AA Dordrecht, The Netherlands.

www.springer.com

Printed on acid-free paper

All Rights Reserved
© 2008 Springer Science+Business Media B.V.
No part of this work may be reproduced, stored in a retrieval system, or transmitted
in any form or by any means, electronic, mechanical, photocopying, microfilming,
recording or otherwise, without written permission from the Publisher, with the exception
of any material supplied specifically for the purpose of being entered and executed on
a computer system, for exclusive use by the purchaser of the work.

CONTENTS

Preface ix

Acknowledgement xi

1. National Nuclear Centre Solving Radiation Safety Problems
   in Kazakhstan 1
   K.K. Kadyrzhanov

2. Radioactive Particles Released from Different Nuclear Sources:
   With Focus on Nuclear Weapons Tests 7
   B. Salbu

3. Phenomenology of Underground Nuclear Explosions in Rock Salt 19
   Yu.V. Dubasov

4. Radiation and Hydro-Chemical Investigation and Monitoring
   of Transboundary Rivers of Kazakhstan 35
   K.K. Kadyrzhanov, H.D. Passell, V.P. Solodukhin, S. Khazhekber,
   V.L. Poznyak, E.E. Chernykh

5. Tritium in Streams, Well Waters and Atomic Lakes
   at the Semi-Palatinsk Nuclear Test Site: Present Status
   and Future Perspectives 41
   P.I. Mitchell, L. León Vintró, J.G. Howlett, M. Burkitbayev,
   N.D. Priest, Yu.G. Strilchuk

6. Safe Management of Residues from Former Uranium Mining
   and Milling Activities in Central Asian IAEA Regional Technical
   Cooperation Project 61
   P.W. Waggitt

7. Rehabilitation of Uranium Mines in Northern Tajikistan 69
   M.M. Yunusov, Z.A. Razikov, N.I. Bezzubov, Kh.I. Tilloboev

8. A Rational Approach to Bridging the Nuclear Technology
   Usage and Nuclear Education Gap 77
   S.N. Bakhtiar

9. Strontium-90 Contamination Within the Semipalatinsk
Nuclear Test Site: Results of Semirad1 and Semirad2 Projects –
Contamination Levels and Projected Doses to Local Populations 87
*N.D. Priest, Y. Kuyanova, P. Pohl, M. Burkitbayev, P.I. Mitchell,
L. León Vintró, Y.G. Strilchuk, S.N. Lukashenko*

10. The Joint Convention on the Safety of Spent Fuel Management
and on the Safety of Radioactive Waste Management:
An Instrument to Achieve a Global Safety 107
P. Risoluti

11. Reduction of Risks from Lira Underground Nuclear Facilities
at Karachaganak Oil-and-Gas Complex 115
T.I. Ageyeva, A.Zh. Tuleushev, V.V. Podenezhko

12. The Net Effect of the Armenian Nuclear Power Plant
on the Environment and Population Compared to the Background
from Global Radioactive Fallout 125
K. Pyuskyulyan, V. Atoyan, V. Arakelyan, A. Saghatelyan

13. Assessment of Risks and Possible Ecological and Economic
Damages from Large-Scale Natural and Man-Induced Catastrophes
in Ecology-Hazard Regions of Central Asia and the Caucasus 133
*A.N. Valyaev, S.V. Kazakov, A.A. Shamaeva, O.V. Stepanets,
H.D. Passell, V.P. Solodukhin, V.A. Petrov, G.M. Aleksanyan,
D.I. Aitmatova, R.F. Mamedov, M.S. Chkhartishvili*

14. Distribution of Natural and Technogenic Radioactivity in Soil
Samples from Foothill and Mountain Areas in Central Tajikistan 151
A.A. Juraev, An.A. Dzuraev, T. Davlatshoev, H.D. Passell

15. Quantitative Assessment of the Man-Induced Uranium in the Tail
Disposal of Kara-Balta Mining Plant 167
*V.M. Alekhina, I.A. Vasiliev, S. Idrisova, S. Mamatibraimov,
G.M. Tolstikhin*

16. Study of the ^{222}Rn Distribution in Air Above the Tail Disposal Site at KBMP
(Kyrgyz Republic) and Its Possible Transfer to Adjacent Territories 173
*I.A. Vasiliev, V.M. Alekhina, S. Mamatibraimov, O.I. Starodumov,
S. Idrisova, D.A. Ivanenko*

17. Study of Transborder Contamination of the Syr-Darya and Amu Darya
Rivers and Their Inflows 181
*B.S. Yuldashev, H.D. Passell, U.S. Salikhbaev, R.I. Radyuk, A.A. Kist,
S.V. Artemov, G.A. Radyuk, E.A. Zaparov, E.A. Danilova, A.A. Zuravlev,
V.S. Vasileva, E.E. Lespukh*

18. The Navruz Project: Cooperative, Transboundary Monitoring, Data Sharing and Modeling of Water Resources in Central Asia.................... 191
    H.D. Passell, V. Solodukhin, S. Khazekhber, V.L. Pozniak, I.A. Vasiliev, V. Alekhina, A. Djuraev, U.S. Salikhbaev, R.I. Radyuk, D. Suozzi, D.S. Barber

19. Several Approaches to the Solution of Water Contamination Problems in Transboundary Rivers Crossing the Territory of Armenia................ 201
    G.M. Aleksanyan, A.N. Valyaev, K.I. Pyuskyulyan

20. Study of the Ecological State of Rivers Kura, Araks and Samur on Azerbaijan Territory... 213
    R.F. Mamedov, G. Magerramov

21. Ecological Considerations Related to Uranium Exploration and Production.. 219
    P.G. Kayukov

22. Joint Norwegian and Kazakh Fieldwork in Kurday Mining Site, Kazakhstan, 2006.. 225
    G. Strømman, B.O. Rosseland, J. Øvergaard, M. Burkitbayev, I.A. Shishkov, B. Salbu

List of Contributors 233

PREFACE

There is a significant number of nuclear and radiological sources in Central Asia, which have contributed, are still contributing, or have the potential to contribute to radioactive contamination in the future. Key sources and contaminated sites of concern are:

The nuclear weapons tests performed at the Semipalatinsk Test Site (STS) in Kazakhstan during 1949–1989. A total of 456 nuclear weapons tests have been performed in the atmosphere (86), above and at ground surface (30) and underground (340) accompanied by radioactive plumes reaching far out of the test site.

Safety trials at STS, where radioactive sources were spread by conventional explosives.

Peaceful nuclear explosions (PNEs) within STS and outside STS in Kazakhstan, producing crater lakes (e.g., Tel'kem I and Tel'kem II), waste storage facilities (e.g., LIRA) etc.

Technologically enhanced levels of naturally occurring radionuclides (TENORM) due to U mining and tailing. As a legacy of the cold war and the nuclear weapon programme in the former USSR, thousands of square kilometers in the Central Asia countries are contaminated.

Large amounts of scale from the oil and gas industries contain sufficient amounts of TENORM.

Nuclear reactors, to be decommissioned or still in operation.

Storage of spent nuclear fuel and other radioactive wastes.

In the characterization of nuclear risks, the risks are estimated by integrating the results of the hazard identification, the effects assessment and the exposure assessment. A hazard is defined as a factor or exposure that may adversely affect the health; it is basically a source of danger. Hazard is a qualitative term expressing the potential of a source to harm the health of individuals or populations, if the exposure level is sufficiently high and/or if other conditions apply. A risk is defined as the probability that an event will occur having an unfavorable outcome (probability of an event multiplied with the consequences of that event).

In the assessment of nuclear risks, the probabilities of accidents and their consequences are assessed. The analysis includes evaluation of the sources and possible accidental scenarios, ecosystem transfer, biological effects as well as social and economic consequences following the event. Sources may occur stationary (point sources), temporally (labile sources), and outside an individual country. In general, the larger the inventory of radionuclides, the greater the hazard, unless specific safety precautions are taken. In most cases, the sources are known and inventories (Becquerels, Bq) are well established, while in other cases (e.g., old waste storage facilities) the information may be less complete or lacking. When it comes to unforeseen events such as sabotage and terrorisms, neither the source (inventory) nor the localities are known. However, risk assessments and priorities of key sources of concern can be utilized to introduce more safety measures and to build up a relevant emergency response. Thus, environmental impact and risk assessments can form the basis for practical policy-making, such as authorization of industrial releases, interventions within highly contaminated areas, countermeasures (e.g., food restriction), clean-up strategy and remediation of contaminated areas, as well as the updating of legislations and laws associated with the radiation protection of man and the environment. As several countries in Central Asia are facing nuclear risks from a lot of sources and because the contamination is transboundary, a regional and international co-operation within this field seems highly relevant.

The NATO Environmental Security Panel has recognized key hazards of concern: Natural Hazards, Human Induced Environmental Hazards, and the Degradation of the World's Natural Resource Base. Thus, nuclear and radiological risks are well within the scope of NATO ESP. Nuclear and radiological risks and potential transboundary contamination represent an international challenge, and are highly relevant for Central Asia. Therefore, the NATO Advanced Research Workshop (ARW): "Nuclear Risks in Central Asia" was organized as a top-down initiative, in collaboration with the selected co-directors K.K. Kadyrzhanov, Institute of Nuclear Physics NNC RK, Almaty, Kazakhstan and B. Salbu, Norwegian University of Life Sciences.

The objectives of the ARW was to
- Evaluate radiological consequences and risks associated with nuclear weapon tests, peaceful nuclear explosions, uranium mining and milling industry, as well as scale from oil and gas production in Central Asia countries.
- Gather scientists with top competence and experience from contaminated sites in the Central-Asia countries to discuss radiological consequences and risks associated with historical, present and future potential releases of radionuclides in the region.
- Create a meeting place to enforce ongoing collaborations, to establish new relationships, to establish links between ongoing projects and to establish new joint international projects.

The meeting fulfilled the objectives of the ARW, and more than 80 participants attended the meeting. The present book summarizes most of the presentations given at the workshop.

The co-directors will thank the Institute of Nuclear Physics NNC RK in Almaty for the successful organization of the meeting, including an excellent reception dinner for the participants. We are especially indebted to Elvira Chernykh, INPh, for the administrative support.

The organizers are also pleased that the NATO Sciences Committee decided to initiate the NATO ARW "Nuclear Risks in Central Asia", and hope that the publication of the results will find its way to the many scientists, regulators and other interested parties in this field.

On behalf of the NATO ARW co-directors
Brit Salbu

ACKNOWLEDGEMENT

Many people contributed to the success of the NATO Advanced Research Workshop (ARW) on "Nuclear Risks in Central Asia" held June 2006 and organized by the co-directors K.K. Kadyrzhanov, Institute of Nuclear Physics NNC RK, Almaty, Kazakhstan and B. Salbu, Norwegian University of Life Sciences (UMB). The editors are grateful to Elvira Chernykh, INPh, for collecting papers, to many of the participants, as well Dr. Ole-Christian Lind, UMB, for assistance with proof reading of the papers and to Signe Dahl, UMB, for lay-out. We are also indebted to the Springer for fruitful assistance with respect to the publication of papers presented at the NATO ARW "Nuclear Risks in Central Asia"

Lindis Skipperud and Brit Salbu, editors.

CHAPTER 1

NATIONAL NUCLEAR CENTRE SOLVING RADIATION SAFETY PROBLEMS IN KAZAKHSTAN

K.K. KADYRZHANOV
The Institute of Nuclear Physics NNC RK, Almaty, Kazakhstan

Abstract: Kazakhstan has inherited from USSR a series of different radiation-hazardous objects and contaminated sites in all parts of the country, as a legacy of the cold war. The paper presents the key sources representing the radiation risks of concern; nuclear weapons tests, PNEs, uranium mining and tailing, as well as scale from oil and gas production.

Keywords: nuclear sources, nuclear weapons tests, PNE, uranium mining and tailing, scale, Kazakhstan

1. Introduction

Kazakhstan is a country that voluntarily has refrained from possessing nuclear weapon; and is a country with unique nuclear heritage from the former USSR where still many people recognize the words "*nuclear physics*" as "*NUCLEAR HAZARD*". Kazakhstan has inherited from USSR a huge variety of radiation-hazardous sites and objects located in all parts of the country (Fig. 1).

2. Sources of Radionuclide Contamination

Kazakhstan has presently four research nuclear reactors in Kurchatov and Alatau and one power nuclear reactor in Aktau city. For several decades in the last century various nuclear tests were performed at the territory of Kazakhstan. Most of the explosions were performed for military purposes, but there were also so-called peaceful nuclear explosions (PNE) related to industrial activities, e.g., Azgir, LIRA and others.

The situation at Semipalatinsk Nuclear Test Site (SNTS) attracts particular attention. SNTS territory covers about 18,500 km^2 and includes five test sites where differrent types of explosions were performed. At Opytnoye Pole site air and surface tests,

Figure 1. Radiation-hazardous sites in Kazakhstan

including nuclear and hydro-nuclear explosions were performed. At other sites were performed underground explosions including both camouflaged and excavational ones. Each type of explosion can be characterized by the amount, composition and the area contaminated by radioactive fallouts from the explosions; the localization of radionuclides at various areas and the form of radionuclides in soil (Fig. 2).

In Kazakhstan there are more than 100 sites with radioactive tailings formed due to uranium geological prospecting, exploration and mining. Area covered by such tailings comprises, according to preliminary estimations, up to 35–40 km^2 with a total mass of about 400,000 t. One of the urgent problems in Kazakhstan is the accumulation of radioactive waste from the prospecting, extraction and initial processing of hydrocarbons. Radioactive wastes are accumulated in so high amounts and concentrations that may represent considerable hazard to oil-field personnel and local population.

Lots of radiation-hazardous sites and objects in Kazakhstan have contributed to extremely high level of public concerns and radiophobia that, in many cases, is supported by unavailability of reliable information. As mentioned above, many people in the country associate the words "nuclear physics" with nuclear-related consequences and, namely, with the word "hazard". The Institute of Nuclear Physics makes considerable effort to study these problems thoroughly and to normalize the radio-ecological situation.

3. NCC Activities

Since early 1990s the NNC RK has been running large-scale over-all radioecological investigations at SNTS. The following investigations have been performed:

- Characterization of SNTS sites with respect to contamination and the radiation hazard, and a vast databank (N > 104) has been establish on the distribution of artificial radionuclides Cs^{137}, Sr^{90}, Pu^{238} и $Pu^{239+240}$.

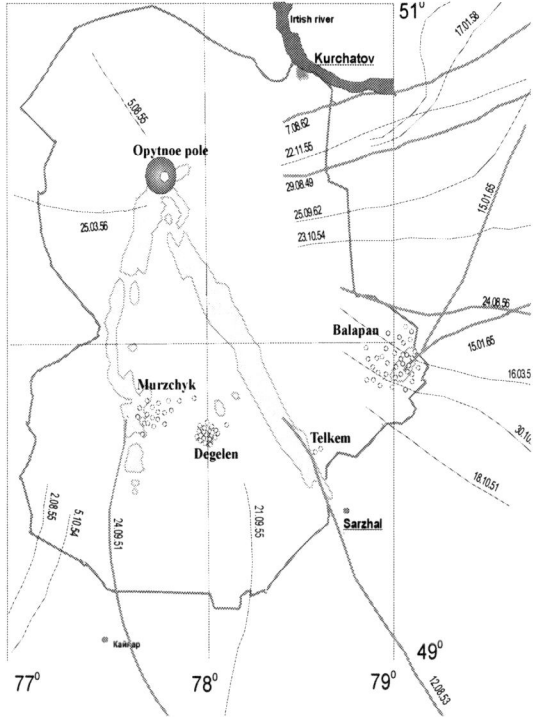

Figure 2. Semipalatinsk Nuclear Test Site

- Investigations of the rehabilitation of contaminated land at SNTS and their transfer for commercial utilization.
- Investigation of contaminated water at SNTS.
- Radiation safety assurance at operation of coal field Kara-Zhyra.

The results show that the primary concern in the future will be the investigation of tritium, the character of tritium and the levels of the environmental contamination, as well as the forecasts of the dynamics of the tritium contamination. Large amounts of heavy hydrogen (deuterium and tritium) were released during nuclear explosions. Almost all heavy hydrogen released to atmosphere has been transferred to ground waters due to isotope exchange processes. Despite of the fact that the increase in the corresponding world average total radiation dose to population is small, releases to the environment of heavy hydrogen is of particular importance because of the role water and hydrogen plays in human life. Thus, further development of technologies and further investigations will be performed.

Works in this direction are planned within the ISTC Project K-1203 "High-Sensitive Ion Technology for Measurements of Low Concentrations of Heavy Hydrogen Isotopes." The objectives of the Project are to develop new technology for the determination of heavy hydrogen isotopic concentrations in hydrogen media accompanied by by the development of a corresponding measuring device, to study the analytical characteristics of the device and to assure its possibilities for use at SNTS. The project should make it possible to achieve sufficiently low sensitivity threshold in determination of tritium (at the level 10–18%) to allow quantitative determination of the tritium background concentrations in the environment.

Less widely known are the nuclear weapons tests performed at other sites: LIRA, Galit and Say-Utes. At LIRA site 700–900 m below surface, six cavities were created and a volume of 45,000–66,000 m^3 was designed for storage of gas condensate from Karachaganak oil-and-gas field (KOGF). Four of them are still partially filled with gas condensate (for 10–75%). At the Galit site in a salt rock strata 17 explosions were performed for the same purpose. At the site Say-Utes were performed three more explosions. In addition, six explosions were performed within programs associated with geological investigations of the Earth crust; "Meridian", "Region", and "Batolit".

The Karachaganak oil-and-gas-condensate field is included in the project "Over-all investigation and monitoring of LIRA facilities". Complex radioecological investigations have been performed at the LIRA facilities, adjacent territories and inhabited localities. Obtained data were used in establishing a monitoring system and radiation safety assurance system at KPGF. Current tasks at LIRA site are:

- Determination of the legal status of the LIRA facilities
- Feasibility studies and technology development for sealing off the nuclear cavities.

The work will be primary directed to the lowering of radiation risks associated with long-term operation of KOGF (Fig. 3).

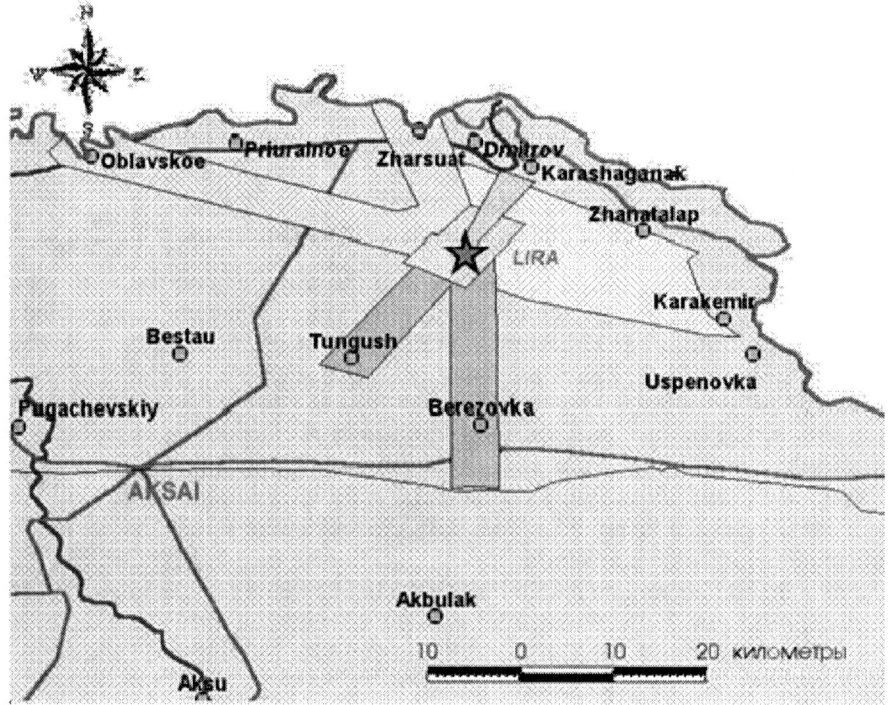

Figure 3. Involvement in Karachaganak Project

4. Naturally Occurring Radionuclides

Last years brought people worldwide to understanding that radiation safety is not the domain of nuclear test sites only; it is an important issue in relation to oil extraction, mining and uranium mining, and water supply industries. This list is also relevant for Kazakhstan. Many years of systematic studies showed that in modern Kazakhstan the radiation risks associated with natural radionuclides is of primary concern.

Investigations of the radiation situation performed in 1992–1995 on territories associated with three oil companies in Western Kazakhstan revealed 267 areas with radioactive contamination due to accumulation and redistribution of natural radionuclides. Some parts of the mining equipment were so highly contaminated with radioactive elements that the radiation doserates exceed 10 mR/h. Net concentrations of lead-210 and radium-226 in the sludge of oil mining equipments could be as high as 100,000 Bq/kg. Oil extraction companies estimate the amount of radioactively contaminated pipes to be about 500,000 t in Mangystau oblast only. The Institute of Nuclear Physics together with the Asia-Clean company has designed and put into operation for the MangystauMunaiGas (Zhetybai village) company facilities clean-up of oil-well tubing, oil and gas separators, technological equipment of water supply companies. Production capacity of the facility comprises 100 t/day at a clean-up rate of 100%.

Kazakhstani land is particularly reach with respect to uranium. Former Soviet Union extracted about 80,000 t of uranium from Kazakhstan, and currently only five uranium mining enterprises are in operation. These enterprises employ highly qualifies personnel. Compliance with regulations and norms at these enterprises assure safe radiation levels both for the personnel and public. One should keep in mind that wastes accumulated at these enterprises during decades of USSR military production require presently huge investments to assure safe storage and disposal of the waste.

Even higher hazards are associated with abandoned objects in the country – there are more than a hundred of such objects of concern. One example is a uranium mining site in the vicinity of the recreation sanatorium Zharkent-Arasan. Following a survey hazards from two mines and an ore dump were recognized: the gamma doserates reached 1,000 µR/h and the concentrations of natural radionuclides within the uranium series were registered at levels up to 10,000–20,000 Bq/kg. The radiation survey resulted in the development and implementation of countermeasures to lower the radiation risks and, consequently, public concerns.

5. Conclusions

Experience gained by the NNC RK during the years based on radioecological and environmental investigations sets us in the position to formulate the main strategic objectives as follows:
- Transition from research ecology to engineering ecology
- Maximize the lowering of radiation risks and transfer of contaminated lands into commercial utilization.

CHAPTER 2

RADIOACTIVE PARTICLES RELEASED FROM DIFFERENT NUCLEAR SOURCES: WITH FOCUS ON NUCLEAR WEAPONS TESTS

B. SALBU
Isotope Laboratory, Department of Plant and Environmental Sciences, Norwegian University of Life Sciences, P.O. Box 5003, N-1432 Aas, Norway

Abstract: Radionuclides released from a source may be present in different physico-chemical forms, varying from low molecular mass (LMM) species, colloids and pseudo-colloids, to particles and fragments. Following releases from severe nuclear events such as nuclear weapon tests, weapon grade materials such as U and Pu are predominantly transported and deposited as radioactive particles or fragments. These entities can carry substantial amounts of refractory fission products, activation products and transuranics. Similarly, radioactive particles are also released following conventional explosions of nuclear weapons or depleted uranium ammunitions, or following nuclear reactor accidents involving explosions or fires. Finally, radioactive particles and colloids are present in effluents from reprocessing facilities and civil reactors entering the environment, and radioactive particles are observed in sediments in the close vicinity of radioactive waste dumped at sea. Thus, releases of radioactive particles containing refractory radionuclides should also be expected following severe nuclear events in the future.

To perform long-term impact assessments for organisms in radioactive contaminated areas by contamination, information on the source term, i.e. activity concentrations, isotopic ratios as well as the radionuclide speciation is essential. If areas are contaminated with radioactive particles, particle characteristics such as the particle size distribution, crystallographic structures and oxidation states are important for assessing particle weathering rates and the subsequent mobilisation and biological uptake of associated radionuclides. Thus, advanced solid-state speciation techniques such as electron microscopy combined with synchrotron radiation X-ray microscopic techniques are needed in radioecology. Many years of research on radioactive particles from different sources has demonstrated that the activity concentrations and the isotopic ratios are source dependant, while particle characteristics such as particle size distribution, crystallographic structures and oxidation states for matrix elements also reflect the release scenario, dispersion processes and deposition conditions.

Keywords: radioactive particles, nuclear weapon tests, Semipalatinsk test site

1. Introduction

Radionuclides in the environment can be present in different physico-chemical forms such as low molecular mass (LMM) species, colloids, pseudocolloids and particles [1]. Radioactive particles in the environment are defined as localised aggregates of radioactive atoms that give rise to an inhomogeneous distribution of radionuclides significantly different from that of the matrix background [2]. Particles in the aquatic environment are defined as entities having diameters larger than 0.45 µm, i.e. will settle due to gravity. Particles larges than 1 mm are often referred to as fragments. Colloids are defined as localised heterogeneities ranging in size from 1–10 nm, while or pseudocolloids or polymers may range from 10 nm to 0.45 µm. In air, these entities are referred as aerosols. The LMM fraction refers to species less than 1 nm or molecular mass less than about 1 kDa [1]. The presence of radioactive species ranging from colloid to fragments can easily be identified by autoradiography, reflecting their inhomogeneous distributions in soils, sediments and waters [1].

Radioactive particles containing refractory radionuclides such as U and Pu are formed due to critical or subcritical destruction of weapon-grade or fuel matrices (e.g., explosions, fires, corrosion processes), cluster formation, condensation processes or interactions with available particle surfaces during release and dispersion. A significant fraction of refractory radionuclides released during high temperature nuclear events, such as nuclear weapons tests (e.g., Maralinga, Mururoa, Nevada, Marshall Islands) and nuclear explosions (e.g., Chernobyl), and fires in nuclear reactors (e.g., Windscale) is present as particles [3–10]. During high temperature and high pressure may liquefy the materials, volatiles will escape, while refractory fission and activation products will remain when droplets are solidified. The presence of construction materials and Oxygen during the events is also critical for the products formed, i.e. the formation of metal-metal crystalline structures such as U–Zr or carbide compounds (Zr from Zirkaloy in fuel elements, C from Carbon moderators in reactors) as well as the oxidation state of the matrices.

Radioactive particles have also been released under low temperature and normal pressure conditions such as atmospheric emission during normal operations (Windscale, early 1950s), accidental discharges (e.g., Krashnoyarsk, Russia) or authorised effluents from nuclear installations (Sellafield, UK; La Hague, France) and have been observed in sediments contaminated from dumped radioactive material [11–15]. Due to insufficient filtering of air emissions or insufficient clean-up of effluents U and Pu particles containing a series of fission and activation products are released into the environment. However, characteristics of such particles are different from those released during high temperature and pressure events. Characteristics such as the activity concentration and isotope ratios of refractory radionuclides, in particular of matrix elements such as U and Pu, will reflect the emitting source. Furthermore, release conditions (high temperature and pressure, in presence of air (Oxygen) will influence the activity concentrations of volatiles (e.g., ^{137}Cs, ^{90}Sr) as well as particle size distributions, crystallographic structures and oxidation states.

2. Characterisation of Radioactive Particles

Radioactive particles released from a source are inhomogeneously deposited and distributed in the environment. Such localised heterogeneities of radionuclides represent a radioanalytical challenge; samples collected may not be representative of the bulk [16], sample dissolution may be incomplete [17] and consequently contamination inventories may be underestimated.

To obtain information on radioactive particles, particles must be extracted from its surroundings and solid-state speciation techniques should be applied, prior to any dissolution of particles. By tedious sub-sampling and application of autoradiography as well as non-destructive gamma-, beta- or alpha-spectrometry, the size, and structure of individual particles can characterised by scanning electron microscopy (SEM). Using backscattered electrons, information on structure as well as the distribution of high atomic number elements on particle surfaces can be obtained, while in transmission electron microscopy, electron dense colloidal sized structures can be identified. From X-ray mapping, the 2D distribution of individual elements associated with particle surface can be obtained, while semi-quantitative elemental analysis at specific particle sites can be obtained by X-ray microanalysis [18]. Using synchrotron radiation (SR) based X-ray microscopic techniques, information on 2D or 3D distribution of elements within particles (X-ray absorption/fluorescence), crystallographic structures (μ-X-ray diffraction) and oxidation states of matrix elements (μ-X-ray absorption near edge spectroscopy, μ-XANES) can be obtained [19].

3. Sources Contributing to the Release of Radioactive Particles into the Environment

According to UNSCEAR [20] more than 2,000 nuclear weapons tests have been performed globally, in the atmosphere, at ground, under ground, and under water. This is the key source contributing to radioactive contamination globally and has also contributed to the release of radioactive particle. In addition to nuclear weapons tests, there is a series of other sources that have contributed to the release of radioactive particles into the environment.

Radioactive particles have also been released during nuclear accidents. During the fire in Pile No 1 at Windscale, UK (former Sellafield) in 1957, radioactive particles varying in size within 20–500 μm were observed up 4 km from the site [13]. Corroded radioactive particles (about 20 kg U) were also released prior to the fire via the stack of the air-cooled reactor [11]. U fuel particles, several hundred microns sized, had flake-like structures [9] and were inert towards leaching [17]. Following the Chernobyl accident, radioactive particles varying in composition, size, shape, structures and colours have been identified varying from compact small-sized crystalline single particles to large amorphous aggregates [21, 22]. Fragments and large particles settled close to the site, while small-sized particles were transported over large distances, and were identified in Scandinavia more than 2,000 km from the site [23]. Based on synchrotron-radiation X-ray micro-techniques, inert fuel particles with a core of UO_2 with surface layers of U–C or U–Zr were released during the initial explosion. In contrast, more soluble fuel particles with a UO_2 core and surface layers of oxidised uranium were released during the fire [19]. The particle weathering constant was low for particles released during the explosion and high for particles released during the fire.

Radioactive particles have been released due to accidents with nuclear devices. Due to the re-entry of the reactor driven Soviet satellite Cosmos 954 in 1978, radioactive particles ranging from submicrons to fragments contaminated large areas in Canada [24]. Due to the US B52 aircraft accident at Palomares, Spain in 1966, where two thermonuclear weapons were conventially detonated the surrounding area was contaminated with submicron to millimeter particles containing ^{239}Pu and ^{235}U [25]. Based on synchrotron radiation microtechniques, the Pu-U particles were oxidised [26]. Due to the US B52 aircraft accident at Thule, Greenland, 1968, where four thermonuclear weapons were conventially detonated the sediments were contaminated with submicron to about 2 mm particles containing ^{239}Pu and ^{235}U. Based on synchrotron radiation microtechniques, the Pu-U particles were found to be oxidised [27, 28].

Radioactive particles have been released from European and Russian reprocessing plants. Due to discharges from Sellafield, UK, into the Irish Sea, sediments have been contaminated with particles containing U [29]. Due to accidental releases at Dounreay, UK, in the 1960s, U fuel particles containing a series of fission products (MBq) are still collected at nearby beaches [unpublished]. No information on particles released from La Hague, France, is available, but a major fraction of radionuclides in the effluent from La Hague during normal operation are associated with particles and colloids [12]. Radioactive particles have also been released into the Techa River from Mayak PA and into the Yenisey River from Krasnoyarsk Mining and Chemical Combine [30, 14].

Large amounts of radioactive and nuclear waste, including six submarined with fuelled reactors, have been dumped in the Kara Sea and in the fjords of Novaya Zemlya during 1959–1991 [24]. Radioactive particles, in particular crud particles, in sediments have also been identified in the close vicinity of radioactive waste dumped in the Abrosimov, Stepovogo and Tsivolky bays at Novaya Zemlya [15].

Depleted uranium ammunitions were applied during the Gulf war (1992) and the 1999 Balkan conflict. Following detonations, DU particles ranging from submicrons to several hundred micrometers, mostly in the respiratory fraction, have been observed in sand, soils and in damaged vehicles. U in the DU particles was oxidized to UO2, U3O8 or a mixture of these oxidation states. Following a fire in a DU ammunition storage facility, up to millimeter sized particles with U present in oxidation state +5 and +6, have been identified [31–33].

4. Particles Released During Nuclear Weapons Tests

The presence of radioactive particles in the environment released from atmospheric and surface ground nuclear weapons tests is well documented from Nevada Test Site in USA, Marshall Islands, Maralinga in Australia, and Mururoa in French Polynesia. The presence of radioactive particles deposited in several areas within the Semipalatinsk STS in Kazakhstan has been observed during recent years. Particle releases from Loop Nor in China have been observed in Japan, while particles released at Novaya Zemlya are identified in Norway (unpublished). For other test sites (e.g., India, Pakistan) information on particle releases is not available in open literature. A series of peaceful nuclear explosions (PNE) has also been performed in USA and former Soviet Union, including Semipalatinsk test site (STS) in Kazakhstan. In addition, safety trials have been performed, where nuclear material has been dispersed into the environment (e.g., Maralinga and Mururoa sites, Semipalatinsk STS) by conventional explosives.

The need to characterise radioactive particles was early recognised. For relatively large particles released during tests at Nevada Test Site and Marshall Islands, information on activity distribution (autoradiography), particle size distribution (rough estimates), shape and colour (subjective statements), and occasionally density and ferromagnetic properties became available in the 1960s [5]. Furthermore, it was realised that the particle characteristics depended on the device and shot conditions and that the leaching efficiency depended on "the nature of the particle". For other test sites, where particles have been documented, only limited information on individual particles and their characteristics influencing particle weathering and subsequent mobilisation of associated radionuclides is currently available.

4.1. MARSHALL ISLAND

During 1946–1958, the United States performed 65 nuclear weapons tests in the atmosphere, from barges and towers, at surface ground and under water at the Bikini and Enewetak atolls, Republic of Marshall Islands [5, 20]. Hot spots or localised heterogeneities have been identified especially at the Yvonne Island [34], and pure Pu particles up to millimeter size surrounded by coral matrix have been identified. Based on autoradiography, large spherical particles (0.5–1 mm) with uniformly distribution of radionuclides and irregular several millimeter-sized particles with surface contamination have been observed after ground-surface shots. The particle characteristics (size distribution, shape, colour) depended on devices and shot conditions [5]. From high altitude shots Pu was associated with spherical particles, being inert towards leaching with water. From coral-surface bursts Pu particles associated with debris were relatively soluble in water.

4.2. NEVADA TEST SITE, USA

During 1951–1962, 84 atmospheric tests took place, while more than 900 underground test were performed during 1951–1992. Venting took place during a series of the underground tests. Hot spots and localised heterogeneities in sand reflected the presence of radioactive particles [6]. A large variety of fused or partially fused Pu-particles, as well as large agglomerates consisting of individual small particles differing in colour, specific activity, density and magnetic properties have been identified. Similar to the Marshall Island, the particle size distribution depended on device and shot conditions; at high altitudes spherical small-sized dense particles with activity distributed throughout the particles were obtained, while at ground surface large irregular shaped particles with lower density and specific activities were observed [5]. The leaching of gross gamma or gross beta activity from particles depended also on device and shot conditions, i.e. matrix, particle size and type. Beta emitters in air-burst debris were dissolved in 0.1 M HCl.

4.3. MARALINGA, AUSTRALIA

During 1953 1963, nine nuclear weapons tests and several hundred smaller-scale weapons safety trials were conducted by the UK at the Maralinga and Emu sites, Southern Australia. Radioactive Pu and U particles have been identified in Maralinga, due to 12 safety trials where more than 20 kg of Pu and about the same amount of U were dispersed by conventional explosives to altitudes up to 800 m. Due to wind transport, the actinides were dispersed many kilometers from the detonation sites. Pu particles; up to several hundred microns, finely dispersed Pu particles and Pu-contaminated soil particles have been reported [3]. Most of the activity was associated with particles within the 250–500 μm fraction. However, about 5% of the total mass was present as particles with diameters less than 45 μm, and a respiratory fraction (less than 7 μm) was identified. In addition to ^{239}Pu and U, ^{241}Am was also associated with particles. Based on individual large particles with activities from 30 to 5 kBq characterised by gamma spectrometry and proton-induced X-ray emission spectroscopy, Pu and U were localised on particle surfaces. Leaching experiments using a simulated lung fluid demonstrated the presence of inert particles with low solubility.

4.4. MURUROA, FRENCH POLYNESIA

During 1966–1996, 41 atmospheric and 137 underground nuclear weapons tests were performed at the atolls of Mururoa and Fangataufa, French Polynesia. In addition, five surface and ten underground weapons trials using conventional explosives were performed in the Colette region at Mururoa in which about 3.5 kg ^{239}Pu were dispersed in each test. Pu particles with activities 5–30 kBq (up to 1 MBq) ranging from 200 μm to several hundred micrometers have been identified in the Colette region [35]. By sieving samples of coral debris, more than 99% of the mass and more than 95% of the activity were present as particles larger than 250 μm. Activities of ^{241}Am ranged from 0.2 to 5.6 kBq and the ^{239}Pu/^{241}Am ratio ranged from 3–67. Using optical microscopy and X-ray micro-fluorescence for analysis of 200–500 μm particles, large differences in the surface structures could be observed; from glassy relatively smooth compact surfaces to conglomerates of small particles with rough appearances. Leaching experiments using human serum demonstrated the presence of inert particles with low solubility; less than 0.07% of the particle content [35, 4].

4.5. NOVAYA ZEMLYA, RUSSIA

During 1950s–1990, 88 atmospheric, 39 underground and at least 3 underwater nuclear weapons test have been carried out at Novaya Zemlya. Significant contamination has been localised at the three major test areas, and is also linked to some major events; radioactive plume from underwater test (1955) crossing the Koushny peninsula and contaminating an area of several square kilometers, radioactive plumes from surface tests at Chernaya Bay (1957, 1961) contaminating the Southern as well as the northeastern Novaya Zemlya, and fallout from the low altitude atmospheric explosion in 1957 [24]. Three underwater weapons tests (1955, 1957, 1961) and dumping of waste

(1991) took place in the Chernaya Bay. The sediments were contaminated with Plutonium and Americium, and the total inventory has been estimated to 3 TBq [36]. Localised radioactivity and sample heterogeneity reflected the presence of particles. Radioactive particles have been identified in air filters from 1961 and 1962 at several sites in Norway (unpublished).

4.6. SEMIPALATINSK, KAZAKHSTAN

During 1949–1989, a total of 456 nuclear weapons tests have been performed in the atmosphere (86), above and at ground surface (30) and underground (340) at Semipalatinsk Test Site (STS) in Kazakhstan [37]. Tests at Ground Zero resulted in widespread contamination of radionuclides such as actinides and fission products. Individual tests could be traced by the isotopic $^{240}Pu/^{239}Pu$ ratios [38]. More localised contamination is associated with the use of peaceful nuclear explosions (PNE) for civil purposes; especially the Atomic Lake and the Tel'kem I and Tel'kem II freshwater crater (Fig. 1) lakes. The Tel'kem I and Tel'kem II craters were created by the explosion of one and three 240 T nuclear (plutonium) devices, respectively. The Tel'kem I lake is circular and has an approximate diameter of 50 m, while The Tel'kem II lake is elliptical, being about 130 m long by 45 m wide. The depths of the lakes are 7–10 m. According to IAEA [37], significant levels of actinide contamination, in the form of large particles or fragments could be observed. Pu associated with particles or fragments with activity levels exceeding 50 kBq kg^{-1} have also been reported [39].

Figure 1. Tel'kem II freshwater crater lakes (130 m long, 45 m wide, 7–10 m depth) produced by the simultaneous detonation of three 0.24 kt fission devices. The crater rim rises up to 25 m above the surface of the water [40]

Autoradiography of soil samples from the Tel'kem II cratering nuclear explosion, and of soil and melted rock/concrete samples from Ground Zero have revealed (Fig. 2) the presence of numerous heterogeneities [40]. Using scanning electron microscopy

Figure 2. Autoradiography (P imaging) of samples collected at Tel'kem II crater reflects the presence of radioactive particles [40]

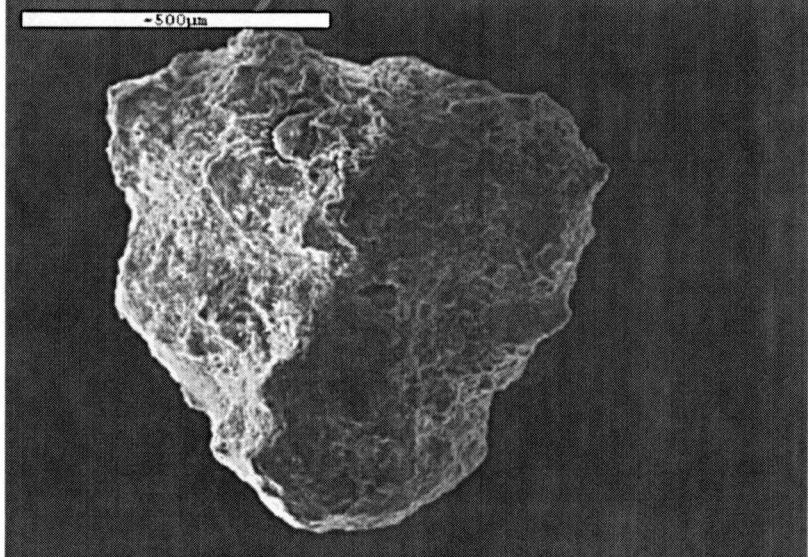

Figure 3. Electron microscopy imaging (SEI mode) of a particle extracted from the Tel'kem II crater. Bar 500 µm [40]

(Fig. 3) the structure of individual particles can be studied [40]. Recent investigations using SEM combined with XRMA, and synchrotron radiation 3D X-ray fluorescence (µ-XRF), demonstrate that the particles formed during the nuclear test have been vitrified, and that the distributions of U and Pu within particles are overlapping particles (unpublished).

5. Conclusions

A major fraction of refractory radionuclides released during serious nuclear events is present as radioactive particles [41]. Several years of research demonstrate that the matrix and refractory radionuclide composition (e.g., burn-up) will reflect the specific releasing source. The release scenarios (e.g., temperature, pressures, redox conditions) will influence particle characteristics such as particle size distribution, crystallographic structures, porosity, and oxidation states. Soil and sediments can act as a sink for deposited particles. Depending on particle characteristics and processes influencing particle weathering and remobilisation of associated radionuclides, contaminated soil/sediments may also act as a potential diffuse source in the future. To assess longterm impact from radioactive particle contamination, the source term should include particle characteristics such as the particle size distribution, crystallographic structures and oxidation states, influencing particle weathering rates and the subsequent mobilisation and biological uptake of associated radionuclides. Furthermore, radioactive particles represent point sources of radiological significance. Following inhalation of respiratory radioactive particles, or surface contamination from the deposition on skin, direct effects relate to internal doses or skin doses. Long-term effects relate to particle weathering and ecosystem transfer of radionuclides. Unless the impact of particles and weathering processes are taken into account, dose and impact assessments of particle- contaminated areas, will suffer from large overall uncertainties. To produce relevant information on source and release dependent particle characteristics, however, advanced analytical microtechniques are needed within radioecology.

References

1. Salbu, B. (2000) Speciation of Radionuclides. *Encyclopaedia Analytical Chemistry.* Wiley, Chichester, 12993–13016.
2. IAEA CRP. (2001) Co-ordinated Research Programme on radioactive particle. Report by an International Advisory Committee, IAEA, Vienna.
3. Cooper, M., Burns, P., Tracy, B., Wilks, M., and Williams, G. (1994) Characterisation of plutonium contamination at the former nuclear weapons testing range at Maralinga in South Australia. *J. Radioanal. Nucl. Chem.* 177, 161–184.
4. Danesi, P.R., De Regge, P., La Rosa, J., Makarewicz, M., Moreno, J., Radecki, Z., and Zeiller, E. (1998) Residual plutonium isotopes and americium in the terrestrial environment at the former nuclear test sites of Mururoa and Fangataufa, Proc. the 7th Intern. Conf, "Low level measurements of actinides and long-lived radionuclides in biological and environmental samples", Salt Lake City, UT.
5. Crocker, G.R., O'Connor, J.D., and Freiling, E.C. (1966) Physical and radiochemical properties of fallout particles. *Health Phys.* 12, 1099–1104.

6. Anspaugh, L.R. and Church, B.W. (1986) Historical estimates of external γ exposure and collective external γ exposure from testing at the Nevada Test Site. 1. Test series through Hardtack II, 1958. *Health Phys.* 51, 35–51.
7. Kuriny, V.D., Ivanov, Y.A., Kashparov, V.A., Loschilov, N.A., Protsak, V.P., Yudin, E.B., Zhurba, M.A., and Parshakov, A.E. (1993) Particle associated Chernobyl fall-out in the local and intermediate zones. *Ann. Nucl. Energy* 20, 415–420.
8. Chamberlain, A.C. and Dunster, H.J. (1958) Deposition of radioactivity in north-west England from the ccident at Windscale. *Nature* 182, 629–630.
9. Salbu, B., Krekling, T., Oughton, D.H., Østby, G., Kashparov, V.A., Brand, T.L., and Day, J.P. (1994) Hot particles in accidental releases from Chernobyl and Windscale Nuclear installations. *Analyst* 119, 125–130.
10. Hamilton, T.F. (2004) Linking legacies of the Cold war to arrival of anthropogenic radionuclides in the ocean through the 20th century, in H.D. Livingston (ed) *Marine Radioactivity*. Elsevier, 23–78.
11. Jakeman, D. (1986) Notes of the level of radioactive contamination in the Sellafield area arising from discharges in the Early 1950s. *UKAEA, AEEW Report* 2104, Atomic Energy Establishment, Winfrith, Dorset.
12. Salbu, B., Bjørnstad, H.E., Sværen, I., Prosser, S.L., Bulman, R.A., Harvey, B.R., and Lovett, M.B. (1993) Size distribution of radionuclides in nuclear fuel reprocessing liquids after mixing with seawater. *Sci. Tot. Environ.* 130/131, 51–63.
13. Chamberlain, A.C. (1987) Environmental impact of particles emitted from Windscale Piles, 1954–57. *Sci. Tot. Environ.* 63, 139–160.
14. Kjerre, L. (2006) Characterization of Radioactive Particles from Krasnoyarsk-26, Russia. Thesis. Norwegian University of Life Sciences, Aas, Norway.
15. Salbu, B., Nikitin, A.I., Strand, P., Christensen G.C., Chumichev, V.B., Lind. B., Fjelldal, H., Bergan, T.D.S., Rudjord, A.L., Sickel, M., Valetova, N.K., and Føyn, L. (1997) Radioactive contamination from dumped nuclear waste in the Kara Sea – results from the Joint Russian-Norwegian expeditions in 1992–94. *Sci. Tot. Environ.* 202, 185–198.
16. Bunzl, K. (1997) Probability of detecting hot particles in environmental samples by sample splitting. *Analyst* 122, 653–656.
17. Oughton, D.H., Salbu, B., Brand, T.L., Day, J.P., and Aarkrog, A. (1993) Under-determination of strontium-90 in soils containing particles of irradiated uranium oxide fuel. *Analyst* 118, 1101–1105.
18. Salbu, B., Krekling, T., and Oughton, D.H. (1998) Characterisation of radioactive particles in the environment. *Analyst* 123, 843–849.
19. Salbu, B., Krekling, T., Lind, O.C., Oughton, D.H., Drakopoulos, M., Simionovici, A., Snigireva, I., Snigirev, A., Weitkamp, T., Adams, F., Janssens, K., and Kashparov, V.A. (2001) High energy X-ray microscopy for characterisation of fuel particles. *Nucl. Instr. and Meth. A* 467, 21, 1249–1252.
20. UNSCEAR. (2000) Sources and effects of ionizing radiation. *The United Nations Scientific Committee on the Effects of Atomic Radiation*. New York.
21. Kashparov, V.A., Ivanov, Y.A., Zvarich, S.I., Protsak, V.P., Khomutinin, Y.V., Kurepin, A.D., and Pazukhin, E.M. (1996) Formation of hot particles during the Chernobyl nuclear power plant accident. *Nucl. Tech.* 114, 246–253.
22. Kashparov, V.A., Oughton, D.H., Protsak, V.P., Zvarisch, S.I., Protsak, V.P., and Levchuk, S.E. (1999) Kinetics of fuel particle weathering and ^{90}Sr mobility in the Chernobyl 30 km exclusion zone. *Hlth Phys.* 76, 251–259.
23. Devell, L., Tovedal, M., Bergstrøm, U., Applegren, A., Chussler, J., and Andersson, L. (1986) Initial observations of fallout from the reactor accident at Chernobyl. *Nature* 321, 817–819.
24. AMAP-Arctic Monitoring and Assessment Programme. (2002) Arctic pollution issues: radioactive contamination. *Report from Norwegian Radiation Protection Authority*. Oesteraas, Norway.
25. Espinosa, A., Aragón, A., Hogdson, A., Stradling, N., and Birchall, A. (1998) Assessment of doses to members of the public in Palomares from inhalation of plutonium and americium. *Radiat. Prot. Dosim.* 79, 1–4.
26. Lind, O.C., Salbu, B., Proost, K., Janssens, K., and Falkenberg, G. (2003) Oxidation state determinations of U and Pu in particles from Thule and Palomares. In: *HASYLAB Annual Report 2002*. Hamburg.
27. Eriksson, Mats. (2002) On Weapons Plutonium in the Arctic environment (Thule, Greenland). Ph.D. thesis, Risø National Laboratory, 3-5-0002, 1–146.
28. Lind, O.C., Salbu, B., Janssens, K., Proost, K., and Dahlgaard, H. (2005) Characterization of uranium and plutonium containing particles originating from the nuclear weapons accident in Thule, Greenland, 1968. *J. Env. Radioact.* 81, 21–32.

29. Jernstrøm, J., Eriksson, M., Osan, J., Tørøk, S., Simon, R., Falkenberg, G., Alsecz, A, and Betti, M. (2004) Non-destructive characterisation of radioactive particles from Irish sea sediment by micro X-ray fluorescence (µ-XRF) and micro X-ray absorption near edge spectroscopy (µ-XANES). *J. Anal. At. Spectrom.* 19, 1428–1433.
30. JNREG-Joint Norwegian-Russian Expert Group for Investigation of Radioactive Contamination in the Northern Areas. (1997) Sources contributing to radioactive contamination of the Techa River and areas surrounding the "Mayak" production association, Urals, Russia. *Report by Norwegian Radiation Protection Authorities,* Østerås, Norway, ISBN 82-993979-6-1, 134.
31. Danesi, P.R., Markowicz, A., Chinea-Cano, E., Burkart, W., Salbu, B., Donohue, D., Ruedenauer, F., Hedberg, M., Vogt, S., Zaharadnik, P., Ciurapinski, A.J. (2003) Depleted uranium particles in selected Kosovo samples. *Env. Radioact.* 64, 143–154.
32. Salbu, B., Janssens, K., Lind, O.C., Proost, K., and Danesi, P.R. (2003) Oxidation states of uranium in DU particles from Kosovo. *J. Env. Radioact.* 64, 163–167.
33. Salbu, B., Janssens, K., Lind, O.C., Proost, K., Gijsels, L., and Danesi, P.R. (2004) Oxidation states of uranium in depleted uranium particles from Kuwait. *J. Env. Radioact.* 78, 125–135.
34. Simon, S., Jenner, T., Graham, J., and Borcher, A. (1995) A comparison of macro- and microscopic measurements of plutonium in contaminated soil from the Republic of the Marshall Islands. *J. Radio anal. Nucl. Chem.* 194, 197–205.
35. IAEA. (1998) The radiological situation at the atolls of Mururoa and Fangataufa. *Report by an International Advisory Committee*, IAEA, Vienna.
36. Smith, J.N., Ellis, K.M., Naes, K., Dahle, S., and Matisov, D. (1995) Sedimentation and mixing rates of radionuclides in Barents Sea sediments off Novaya Zemlya. *Deep Sea Res. II* 42, 1471–1493.
37. IAEA. (1998) Radiological conditions at the Semipalatinsk test site, Kazakhstan: preliminary assessment and recommendations for further research. *International Atomic Energy Agency, Radiological Assessment Report*, Vienna.
38. Yamamoto, M., Tsumara, A., Katayama, Y., and Tsukatani, T. (1996) Plutonium isotopic composition in soil from the former Semipalatinsk nuclear test site. *Radiochimica Acta* 72, 209–215.
39. Dubasov, U.V., Krivohatskii, A.S., Kharitonov, K.V., and Ghorin, V.V. (1994) Radioactive contamination of the Semipalatinsk province ground and adjacent territories of the region after atmospheric nuclear tests in 1949–1962. Remediation and restoration of radioactively-contaminated sites in Europe. *Proceedings of the Int. Symp., Antwerp 1993.* Doc. XI-5027/94. European Commission, Brussels, 25.
40. EU Advance. (2004) Source-specific ecosystem transfer of actinides utilising advanced technologies, *Final Report, Contract FIGE*-2000-00108, Dublin.
41. Salbu, B. and Lind, O.C. (2005) Radioactive particles released from various nuclear sources. *Radioprotection* 40, Suppl. 1, 27–32.

CHAPTER 3

PHENOMENOLOGY OF UNDERGROUND NUCLEAR EXPLOSIONS IN ROCK SALT

YU.V. DUBASOV
RPA "V.G. Khlopin Radium Institute"
28, 2-nd Murinskii av., St-Petersburg, 194021
Russia

Abstract: Results from investigations of numerous cavities made from underground nuclear explosions conducted in rock salt in the USSR are presented. A physical-chemical model of these nuclear explosions that are based on these investigations has been developed. Problems associated with rock salt evaporation and melting, the cavity temperature regime, progress of rock salt cooling and the distribution of radionuclides in a formed cavity have been described by such model. Scientific problems arising during cavities opening and during preparations for operations have been considered. In addition, problems associated with both radionuclide migration with ground waters and contamination of radioactive products stored in the explosion cavities, are considered.

Keywords: radioactive particles, nuclear weapon tests, Semipalatinsk test site

1. Introduction

The first underground nuclear explosion (UNE) in rock salt formation was conducted in 1961 near Karlsbad, New Mexico, USA. Important program on UNE was initiated in the Soviet Union in 1966. The goal of this program was both technology elaboration to obtain transuranium and transplutonium radionuclides in weighable amounts, and the technology of creating cavities as capacities for hydrocarbon raw materials. On the whole, 40 UNEs were conducted in rock salt in the USSR. Most of explosions were conducted in vertical holes, and seven explosions were performed in cavities created by previous explosions. Thus, in the A-2 explosion cavity filled with water, six low-yield explosions were conducted with the aim to develop transplutonium elements obtaining technology. One explosion with a yield of 10 kt was conducted in the A-III explosion cavity.

The goal of this explosion was to develop a technology of multiplied explosions in one cavity, to obtain a man-made transuranium elements deposit, as well as for checking the possibility of seismic signal relaxing (decoupling) during the explosion in a high volume capacity. Explosions in rock salt were conducted at the sites "Galit" near the settlement Bol'shoy Azgir, at the site "Lira" (6 explosions) near Uralsk city, Republic of Kazakhstan, at the site "Vega" near the settlement "Aksaraiskay" in Astrakhan province (15 explosions), and also at the sites "Magistral" and "Sapfir" in Orenburg province (3 explosions), Russian Federation.

The earliest explosions in the USSR were conducted at the "Galit" site. These explosions were investigated most detailed. Therefore, having in mind the phenolmenology of UNE in rock salts, we were guided mostly by using the data obtained at this site, and also by some data published from the Gnome and Salmon explosions which had been conducted in USA.

2. Phases of Nuclear Explosion in Rock Salts

Long-term investigations of the central zones of underground nuclear explosions as well as calculative-theoretical investigations made it possible to create a physical-chemical model of the underground nuclear explosion. It deals with three stages of the explosion transition:

- Chemical bonds destruction and transitions of the first kind and with energy absorption
- Chemical reactions and transitions of the first kind and energy release
- Migration stage.

The first stage includes chemical decomposition of the substances, nuclear charge materials and a part of the surroundings to high-temperature (5,000–10,000 K) vapour gas mixture, and to the volume-up to practically full ionized plasma, adiabatic explosion of cavity gas, rock melting and its heating up to the temperatures, resulting in chemical changes. Thus, at this state of an explosion, significant break of chemical bonds and transfer into ion-atomic state occurs in the central plasma zone of explosion (fire ball), as well as full substance evaporation with a transfer into atomic-molecular state in the evaporation zone, melting of rock with partial decomposition, and volatile and gaseous disintegration product separation, and, finally, rock heating accompanied by crystal-hydrate destruction and volatile products giving up in the zone of thermal effect. These stages are often overlapping in space and time. The duration of the first stage of process in an ideal case is not more than some seconds.

After the descent of a special nuclear explosive device into a hole, stemming complex was mounted above this device, preventing radionuclides to escape from the hole to the day surface. During nuclear explosion huge energy emission occurs in about 0.1 μsec. The values of arising temperature and pressure could be calculated from formulae given below [1]

$$E_{tot} = E_{tot} = 3/2 \frac{6,023 \cdot 10^{23}}{M} \ldots kT, + 7\ 67 \cdot 10^{-15} T^{-4} = 1,25 \cdot 10^8 \rho/MT$$
$$+ 7,67 \cdot 10^{-15}\ T^4\ \text{erg/cm}^3 \tag{1}$$

where ρ = substance density, M = molecular weight of the substance, k = Boltsman constant. The total pressure of an ideal gas at high temperatures is equal to:

$$P_{tot} = 2/3\ E_{kin} + 1/3\ E_{rad} = nkT + 1/3\sigma_R T^4, \qquad (2)$$

where E_{kin} = kinetic energy, E_{rad} = radiation energy, and σ_R = Stephan-Boltsman constant.

For an explosion yield of 100 kt; weight 5 t and density 2.12 g/cm^3 the temperature in the charge volume will be 1.4×10^7 K [1]. Gas pressure and radiation pressure will be 1,200 Mbar and 75 Mbar, respectively. At such high temperatures, the substance of a charge will convert into highly ionized gas, and the thermal wave will spread in the surroundings. After temperature decreases to 300,000 K, the thermal wave will transit into a powerful shocking wave, a rock salt pressure and the system is transferred into evaporated state.

The second phase includes a period and space of cooling substances in the course of adiabatic expansion, thermal- and mass-exchange when in contact with cavity walls and joints' surface. In fact, it starts in some hundreds of fractions of a second with the appearance of non-condensing gases and clusters' formation (100–200 atoms and molecules) of the most refractory elements and compounds, which is thermodynamically possible in a given mixture of elements, and is completed by gas structure stabilization and rock melt crystallization, fixing basic part of explosion radionuclides with partial restoring of the former minerals and formation of new ones. Suitable space-time stage limit is water vapours condensation that is cooling down the cavity gas and rock to 373 K.

As a cavity size increases, adiabatic expansion and cooling of ionized gas of evaporated rock take place. It results first in ions recombination; for Na and Cl atoms first the recombination into NaCl molecules and then to recombination of ionized molecules NaCl$^+$ with electrons. Recombination energy liberated here is transferred to cavity walls, that is, melted rock, which at this period still covers the walls of the cavity.

At the expense of radiance and convective heat exchange the evaporated rock temperature decreases, while this rock is still a fire ball, and NaCl condensation process begins. Latent condensation (evaporation) heat for sodium chloride is $\Delta H = 1$ Mcal/kg (4.19 MJ/kg) and is spent on melting and heating of melted rock. With due regard to the fact that 100–110 t of rock salt are condensed, the total amount of liberated condensation heat comprises a significant part from the total explosion energy.

Under conditions of shock compression, rock salt will be evaporated, if inner energy accumulated as a result of compression, E_{evap}, is higher than 8.9 kcal/kg, and that for melting –3.4 kcal/kg [2]. These values are 2–3 times higher than under ordinary conditions (normal pressure). The shocking wave amplitude, when rock salt evaporation takes place at the front of shocking wave, is P = 0.86 Mbar (86 GPa). When the shock wave amplitude is diminished to 800 kbar pressure, the rock behind the front of shocking wave melts. When the shocking wave amplitude is diminishing to 400–500 kbar, compressed rock is crushed and heated.

Thus, in course of underground nuclear explosions, evaporation and melting of rock at the front of shocking wave take place at first, rock melts also during a relief (decompressing) of compressed rock, transforming into shock melt. Complementary melting of rock (thermally formed melt) by a previously heated shocking wave occurs at the expense of latent heat in evaporated rock condensation, and mixture of these two melt generations results in a final radioactive melt formation.

Formation of sodium chloride vapours, main component of rock salt condensation, can take place in wide time interval. For instance, in case of water penetration from higher horizons into a cavity of small explosion (1 kt), condensation ended in 10 sec, for an explosion of 25 kt of yield condensation time was ~50 sec. In case of contained cavity rock condensation after an explosion of about 60 kt of power, rock condensation ceased in 4–6 min. These assessments were obtained by radionuclide chronometry method. The average melt temperature 100 sec after the A-I explosion was 1,800°C. In the periphery zone of the A-I explosion cavity, the temperature 10–15 sec after the explosion was 1,550–1,600°C; at the boundary of the melted rock with cavity walls the temperature reached 900–1,000°C, which resulted in additional rock melting. Melted rock in the first period after an explosion (1–2 min) contains 58% of the explosion energy that is 2.44 TJ/kt as thermal energy. Specific thermal energy in melted rock is 2.44 MJ/kg.

Solidification time (crystallization) of melts depends on many factors, in particular on water penetration into a cavity. Thus, for a series of explosions at the "Galit" site water penetration and early melt solidification in upper layers were observed for instance 10 min after the A-I explosion. Despite a high specific activity of solidified melt, it was not melted again under the influence of radiation loadings.

As a result of the powerful initial impulse received by the rock, and also of evaporated rock pressure, cavity formation and its subsequent growth takes place. Numerical calculations show that a cavity is compressed after maximum radius is achieved, and then it is expanded again, that is, up to complete stop of the cavity growth, pulsation occurs [3]. For example, the calculated cavity of "Salmon" explosion 0.1 sec after explosion had a maximum radius of 23 m, and 0.25 sec after the explosion the pulsation cavity expansion was finished, and its final radius was 22 m. Because of imperfection in the calculation technique, the theoretical radius appeared to be larger than a real one – 17.4 ± 0.6 m.

Numerical calculations of nuclear explosion cavity growth at the depth of 833 m and at a yield of 5.3 kt, that is, similar to "Salmon" explosion, were made by V.B. Adamsky and his colleagues. These calculations are presented in Fig. 1. It is seen that the cavity reaches its maximum size of 29 m, ~0,15 sec after the explosion, and then the size is reduced. Full stop of the cavity growth occurs ~0.45–0.5 sec after explosion. Compared to the estimate, the cavity radius was 19.5 m, which is 2 m larger than a real one. Estimations for the A-III and A-IV explosions with similar yields were also made. Expected cavity radius for the explosions was ~40 m, solidified melt level was located 20–25 m from the cavity centre. It is evident, that the calculated radius was in good agreement with the experimental value (38–39 m).

Final cavity radius R and volume V_{cav} can be estimated by means of equations suggested by V. Adamsky and colleagues [4]:

$$R_M = 14.5\, W_{KT}^{1/3}\, (1 - \frac{H}{H_0}),\qquad(3)$$

$$V_{cav} = 12800\, (1 - H/2800)^3 \cdot W,\ м^3 \qquad(4)$$

where H_0 = 2,800 m; W = explosion yield, kt; H = depth of charge location.

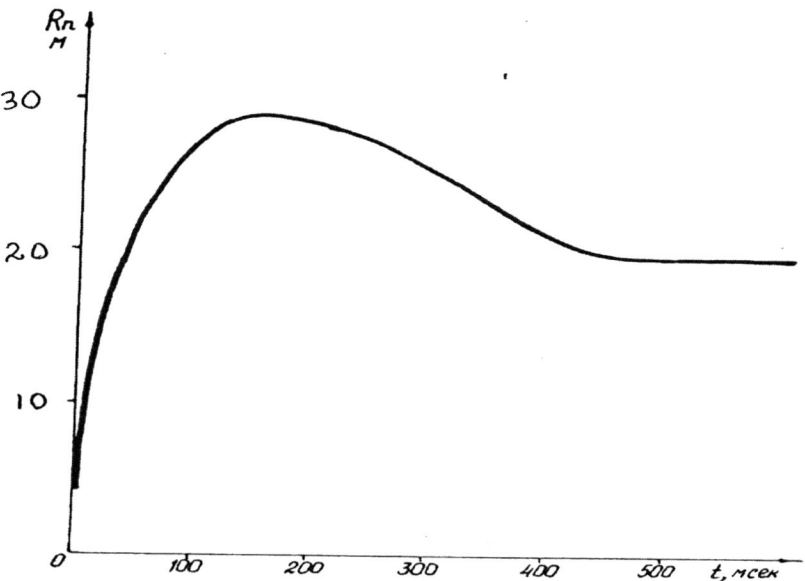

Figure 1. Numerical calculations of the nuclear explosion cavity growth at the depth of 833 m and yield of 5.3 kt

3. Mass of Evaporated and Melted Rock

Masses of evaporated and melted rock are very important characteristics of underground nuclear explosions. They were estimated by US specialists on the basis of hydrodynamic numerical calculations and by our experimental and empirical experiences. According to data from Butkovich [5], the mass of evaporated rock salt is 107 t/kt (24 t/TJ). Our estimations were based on the chemical and radiochemical structure of condensation particles, formed from chemical elements of evaporated rock and charge materials. This estimation was made from results from the investigations of A-I explosion products. According to our data, the evaporated rock mass was 108 ± 15 t/kt (26 ± 4 t/TJ), which is in good agreement with calculated values published earlier [5].

In course of the investigation of the "Gnom" and "Salmon" explosions in USA, experimental data on melted rock mass were obtained. In the "Gnom" explosion, because of a large roof caving, the melt was significantly diluted with crushed rock and completely overlapped with it. Core material obtained as a result of drilling was a mixture of dead rock and radioactive melt with different concentrations. Treatment of obtained materials made it possible for the US investigators to draw a conclusion, that during the explosion ~3,500 t of rock salt were melted, and the specific melt yield for energy liberation unit (kt) was 1,000 t/kt (240 t/TJ) [6]. In a television survey of the "Salmon" explosion, the cavity was geometrically determined from the volume of melt, which allowed the estimation of melted rock mass – 5,000 t. Specific melt mass for this explosion was also 1,000 t/kt (240 t/TJ) [3]. Thus, it was estimated, that as a result of an underground nuclear explosion in rock salt with a nuclear explosion device located in a hole or within a small volume, rock salt melt is formed having specific mass of 1,000 t/kt (240 t/TJ).

4. Temperature in Rock Salt Melt

Experiments of Cormer and co-workers on rock salt shock compression [7] showed that the NaCl crystal melting behind the shocking wave front begins at 54 GPa, and the complete transfer into liquid state is finished at 70 GPa and temperature of 3,700 K. In the pressures interval 54–70 GPa, the temperature of the salt is practically constant because of continuing melting. According to our calculation data, during a nuclear explosion in rock salt, the peak rock temperature at the border of an evaporation zone and melting (R = 2.35 m) is ~6,000 K. In the range of temperatures below 3,200 K, the temperature drop, depending on the distance, can be presented by an analytical equation

$$T(R) = 8{,}2 \cdot 10^4 \cdot R^{-3{,}0} \tag{5}$$

From this, it can be estimated, that rock temperature at the border of zone with the mass of 1,000 t (R = 4.76 m) is 800 K. However, in the process of cavity formation and because melted rock is flowing onto the cavity bottom, heated and melted rock is intensively mixed, and temperature smoothing occurs. We managed to determine this experimentally, when investigating the rock extracted from the A-I explosion cavity. Investigation of core material from radioactive rock in the A-I explosion cavity showed that there were small spherical particles with the size of 0.05–5 mm. They had high specific activity from refractory radionuclides ^{91}Y, $^{95}Zr + ^{95}Nb$, $^{141,144}Ce$ in the order of magnitude of $3 \cdot 10^{12}$ Bq/kg (80 Ci/kg). Spherical form of olivine slag granules formed in the rock salt melt, supported the facts, that these particles initially had a form of silicate melt drops, not mixed with the NaCl melt. Similar granules were found earlier as products from the "Gnom" explosion [8]. According to X-ray-phased and petrographic-mineralogical analysis data, the granules contained structure compounds of olivine type with different ratios of the last compounds of the series: forsterite Mg_2SiO_4-fayalite Fe_2SiO_4, montichellite $CaMgSiO_4$, periclase MgO etc. Their colour changed from white to black depending on their composition. Thus, there were forsterite, montichellite, periclase in white granules, and almost no ferric compounds.

Granules melting temperature was also measured. Depending on the composition, temperature varied in the integral from 1,800°C to 1,560°C. So, white granules with chemical composition similar to rock salt impurity composition were melting at 1,730–1,790°C. Melting temperature of MgO (20%) and forsterite Mg_2SiO_4 (80%) mixture was ~1,800°C [9], melting temperature of montichellite was 1,500°C [10]. Thus, it can be considered that the rock salt melt in the A-I nuclear explosion had a peak temperature not lower than 1,800°C. White granules with the highest melting temperature were found in salt samples from periphery zones and were in a melting state for 1–2 sec, as we determined by using radionuclide chronometry method. Basing on this fact, we came to the conclusion, that in a periphery zone the temperature was lower than

1,800°C, and in this zone the granules solidified quickly. Black granules found in the same zone, had melting (crystallization) temperature of 1,550–1,620°C and were solidified 10–15 sec after the explosion. In the central zone of the melt segment, where granules of interfered composition were found, a temperature between 1,620–1,740°C was kept for about 100 sec after explosion. Other salt impurities could also verify the temperature in peripheral zone.

Thus, in core material samples extracted from the margin zone of the lower cavity, half-spherical pieces of steel with melted edges were present, that is, it was momentary heated up to a temperature about 1,500°C. In the same zone, the compound $MgFe_2O_4$ – magnesioferrite was presented, formed as a result of explosion, while not existing at temperatures higher than 1,500°C. In a zone of melted rock in contact with cavity walls, compounds such as $CaCO_3$, MgO and γ-Ca_2SiO_4, were formed during the explosion, while their existence demonstrates that the temperature was not exceeding 900–1,000°C. Under melted rock there is a layer of recrystallized salt, that is, the salt had been heated to ~800°C. This layer is about 10 cm thick. This layer is assumed to be an initial cavity wall, melted under the influence of the heat emitted by the melt at the bottom of the cavity. Temperature distribution in melted rock segment from the A-I explosion is shown in Fig. 2.

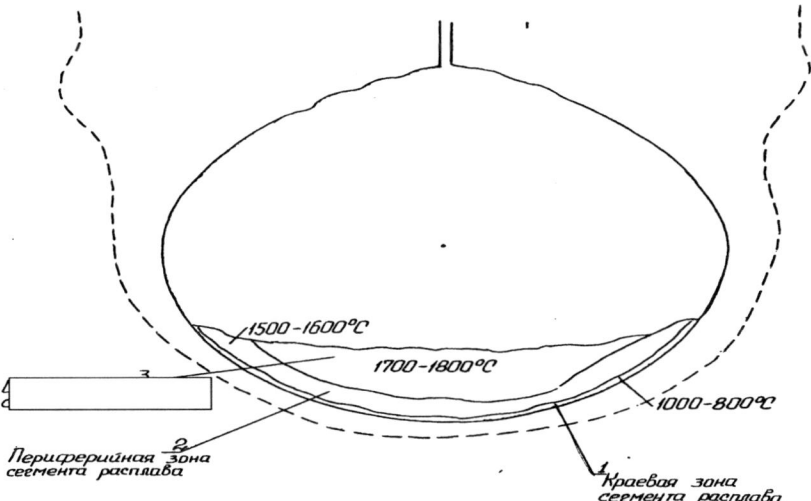

Figure 2. Temperature distribution in a melted rock segment from the A-I explosion

According to Edwards and Holzman [11], the temperature of a melt flowing down to the bottom of "Salmon" explosion crater was in the range from 2,500°C to 800°C, the average temperature of melt collected at the cavity bottom is estimated to 1,400°C [11]. Rawson and co-workers [12], assuming that the formation of compounds such as CaO and Na_2SO_4 during the "Salmon" explosion was a result of the anhydrite $CaSO_4$ reaction with NaCl

$$CaSO_4 + 2NaCl + H_2O = CaO + Na_2SO_4 + 2HCl \qquad (6)$$

supported the fact, that the melt temperature during the first minutes after explosion was about 1,800°C. The rock salt melt temperature in the "Gnom" explosion was also estimated to ~1,800°C [6].

The presence of thenardite Na_2SO_4, found among the A-III-1 explosion products, demonstrates also to a reaction temperature ~1,800°C [6]. Small "balls" of melted iron were also found, as well as larger melted ferric pieces, covered by thick layer of recrystallized rock salt (black colour), under which only ferrous oxide, FeO, was found. Here, we can also speak about a peak temperature not lower than 1,800–1,500°C.

In the A-IV explosion products investigation it was stated, that in melted salt samples of dark colour Na_2SO_4 could be found. This compound did not existing before the explosion. Particles in the form of cubic well-cut crystals of magnetite Fe_3O_4, crystallized, evidently, from Fe_3O_4 melt were also found. Existence of Na_2SO_4 supports that a possible melt heating up to peak temperature was not lower than 1,800°C, and as a fact of magnetite melting – to peak temperature not lower than 1,530°C. Among the explosion products, particles of ferrosilicate slime with melting temperature of 1,220 ± 50°C were present. Particles of this compound were present in a melted state and were transported in a melt influenced by convective currents and turbulence 10–15 min after the explosion.

Thus, experimental data from five explosions, having different yields (from 1 to 60 kt), in rock salt support the facts, that rock salt melts formed during the explosions initially had an average peak temperature not lower than ~1,800°C. In the periphery melt segment zone a temperature of ~900–1,000°C was about constant.

5. Cooling Down of the Cavity

The cavity cooling regime was estimated by numerical methods for explosions, similar to the A-III and A-VI explosions 60 kt yield. Total melt volume was 27,000 m^3, and the density 1.9 g/cm^3. Temperature at the initial time moment t = 0 was estimated to 811°C, not as absolute explosion time, but time of counting "the beginning". According to our data, the melt in the A-I explosion cavity solidified 8–10 min after explosion. It occurred because of fast penetration of water into the cavity. The melt in the Salmon explosion cavity was liquid in its central part for 80–120 days.

Lens (segment) solidification begins from the surface, spreading in increasing the area. The central part of melt segment is solidified latest. The calculations were made with due regard to convection and without convection (the cooling process is slower). In the present paper, convection has been taken into consideration. Total melt crystallization in the lower part of the cavity was completed about 125 days after explosion. Cooling of the central part of melt segment started practically immediately after all melt was present in the bottom cavity. In the course of melt cooling (solidification) the massif is heated to a radius not exceeding 85 m from cavity centre (Fig. 3). Rise of the massif's temperature to more than 100°C is observed at a distance of not more than 20 m from cavity walls. Gas void inside the explosion cavity is supposed to be isothermic, therefore a change of temperature inside the cavity is characterized by temperature dependence at melt surface over time. A cavity cooling process following a lower yield

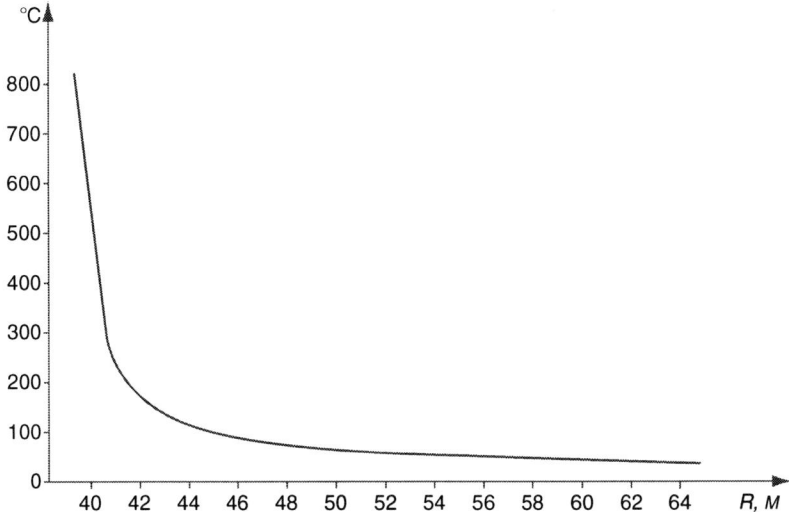

Figure 3. Rock salt temperature outside cavity of A-III-1 explosion (calculation)

explosion occurs faster. Initial moment of cavity cooling would be similar to one considered above, but with sharper temperature drops in the cavity.

Cavity cooling has also been considered following the "Salmon" explosion [12]. The initial energy distribution was as follows: 50% in the explosion cavity, more exactly, in melted salt with a temperature of 1,300°C, the remaining is beyond the limits of the cavity. Salt was kept liquid for at least 1 month; it was ascertained by the radionuclide chronometry method and data on ^{131}I and its daughter radionuclide ^{131m}Xe. In accordance with calculations [12], the final salt crystallization was completed 80-120 days after explosion in a point at a mark 3.9 ± 0.2 m below the upper melting border. About 32 months after the explosion, the "Salmon" explosion cavity temperature was 20°C higher than the natural massif temperature. Four years after the A-III and A-IV cavities opening, the temperature in the cavity was 90–95°C. Calculation of the A-III explosion cavity walls cooling is presented in Fig. 4.

The nuclear explosion energy distribution in thermal E_T and kinetic E_K in a surrounding massif can be expressed by the equation:

$$E_K = E_m - E_T = E_o - E_o(V_o/V_1)^{\gamma-1} - E_T \qquad (7)$$

where E_m is the energy transmitted to the massif, that is, initial explosion energy E_o minus the energy left in the cavity after adiabatic expansion, V_o = initial volume, V_1 = cavity volume, γ = adiabata exponent. Calculations show that thermal energy is [13]:

$$E_T = 0{,}75 E_o [\, 1 - (P_1/P_o)^{\gamma-1}\,], \qquad (8)$$

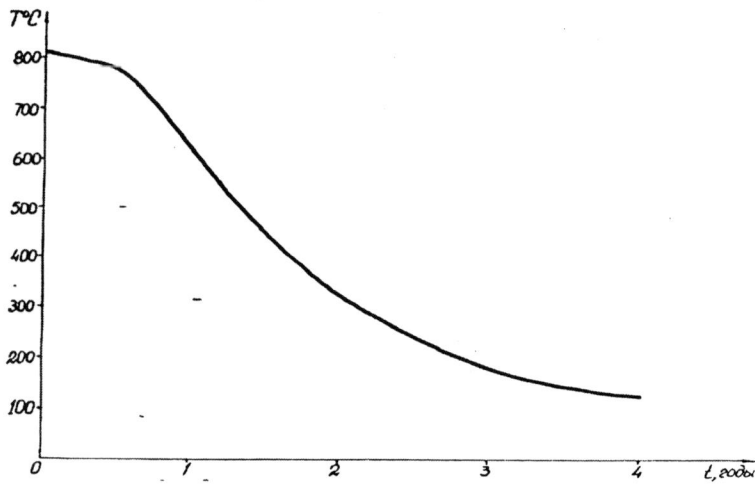

Figure 4. Calculation of A-III explosion cavity walls cooling

where P_o = initial pressure in a cavity, P_1 = threshold pressure at the shocking wave front, whose energy transmitted to the massive is discharged for phase transitions (evaporation, melting) and heating, γ = adiabata exponent. Let us consider which part of the energy is accumulated by melted rock in the initial time moment. Rock salt (NaCl) melting heat is 486 kJ/kg, and the heat capacity at 800°C is 1.09 kJ/kg and increasing with temperature growth. To simplify the calculations it was regarded that average temperature of the rock melt, when it flows down into lower half-sphere of the cavity, is 1,800°C. Multiplying the values obtained for melt masses from 1,000 t/kt, it was obtained that initially the thermal energy in the melt was 2.44 TJ (2.44·10^{12} J), that is 58% of energy for an explosion yield of 1 kt (energy liberation –4.19 TJ). Besides, thermal energy is contained in an isothermal gas and, of course, in cavity walls.

According to the calculations for the "Salmon" explosion [12], the melt having a temperature 1,300°C contained ~50% of explosion energy, the remaining 50% were exponentially distributed beyond the cavity limits with a radius of 17.4 m; ~25% were contained in the first spherical formed layer with thickness of 0.8 m, ~12.5% in the next layer of the same thickness, etc. Four months after explosion, the thermal energy distribution was: 40% above the horizontal plane crossing the cavity centre, 10% in the solidified melt zone and 50% in the lower half-sphere with a radius of ~38 m. After 32 months, the heat was diffused at the distance of nearly 60 m.

At the distance of 70 m from the explosion point, a significant rise in temperature cannot be observed. So, almost 90% of "Salmon" explosion energy was finally transferred into thermal energy and distributed in the cavity and in its vicinity having a radius about 60 m [12]. Besides, about 2% of the energy is expended for seismic wave and activities associated with the cavity expansion.

6. Radiation Effects in Rock Salt

Based on investigation of core materials collected about 0.5 month after the A-I explosion by means of a slope-horizontal hole, dense yellow-orange colour of radioactive rock salt was observed. Later, when core material from the massif located above an explosion cavity was inspected, radioactive salt was found having blue and dark-violet colour. Spectrophotometric investigations of monocrystal plates from the core materials showed yellow salt monocrystals with absorption band with wave length of λ = 476 mμ and conditioned by F-centres existence. The absorption spectrum of blue and dark-violet salt had weakly shown wide band with a maximum in an area with λ = 600 mμ (6,000 Å), conditioned by R_2-centres existence with the absorption maximum 5,960 Å and R-centres with an absorption band at 6,100 Å [14]. It is known, that F-centres are formed during rock salt crystals irradiated by X-rays or other ionizing radiation [15]. Existence time for F-centres is about 10^7 sec (3 months). The salt colour is changed to black, that is, F-centres transform into M-centres. R-centres are formed in salt irradiated at temperatures between 150–220°C or in yellow salt crystals heated up to 200°C, what was also observed under laboratory conditions.

A change in colour into dark-blue of the salt located in the massif above the explosion cavity, supports the facts that irradiation occurred at up-raised temperature. Radionuclide composition of the rock salt samples with dark-blue colour indicated the formation of joints in the rock salt massif after explosion. Along these joints radioactive noble gases, e.g., ^{89,90,91}Kr and ^{131,133,135,137}Xe were moving. As the salt in this area was irradiated at temperatures up to about 150°C, the conclusion was drawn that hot gases were moving along the joints, and also that hot water vapours formed after the first water penetrated into the A-I explosion cavity. Later coloured salt crystals were found among the radioactive salt samples withdrawn to the surface from the A-IV explosion cavity.

7. Migration Phase

The migration phase starts from the moment of cavity opening (unhermetically); it can be overlapped by the second phase and can exist, for the whole period of an explosion zone existence, as an artificial geological body and is characterized by substance recrystallization with a temperature close to geothermal ones.

Nuclear explosion with a yield of 1 kt i.e. is a consequence of instant nuclear fissions of $1.45 \cdot 10^{23}$ nuclei from ^{235}U or ^{239}Pu. In one nucleus decay, two fission fragments are formed, that is, radionuclides with different properties, mass and half-life. The spectrum of fission fragment elements is very wide: from ^{72}Zn to gadolinium ^{159}Gd. In the first minute after fission (explosion) the fission product activity is 20 GCi/kt ($8 \cdot 10^{20}$ Bq/kt). More than 90% of the activity comprises of short-lived radionuclides; radioactive noble gases (RNG) comprise 15%, and together with brome and iodine isotopes they comprise 22%. Initially all radionuclides formed in nuclear fission (fission fragments), residual nuclear fuel, induced radionuclides are together with evaporated rock in gaseous state. Pressure in an explosion cavity for the moment of cavity growth stop can be equal to litostatic pressure of overlying rocks. After the condensation of the great bulk of rock masses, radioactive isotopes of noble gases – krypton and xenon isotopes – are contained in a gaseous phase in the cavity, together with radionuclides of highly volatile compounds, for which condensation occurs at a lower temperature (bromine, iodine, selenium, tellurium, rubidium, antimony, et al.). Radionuclides of noble gases are in principal migrating radionuclides. They can migrate along fractures in rock massif and stemming complex. Process of RNG effluent from central zones can be continued for a long time. The release of RNG depends on both stemming complex reliability and rock salt structures. However, release of gases can be a consequence of a planned technological opening of a cavity through special holes.

Releases from cavities (effluent) into the atmosphere often have two stages. Dry "cold" gas with moisture concentration of 10–20 mg/l and temperature about 20° will leave the cavity first (Table I). Then, after heating of the hole walls by vapour-gaseous mixture, the release of vapour effluent from the cavity was initiated. The most intensive vapour effluent was observed in the E-1, E-3, 2T, 3Tk, 4Tk cavities. In other cases it was mostly less intensive or insignificant. There were no vapour effluents at all from 3T, 8T, 2Tk and 6Tk cavities; and these results indicated that the outlet gases process from cavities was a one-stage process. For technological opening of cavities, pressurized water and gases were expected to be present. But in some cases, excess pressure was not observed; for instance, no excess pressure was observed during drilling of the explosion cavities in 4T, 5T, 7T, 11T (site "Vega") and 2Tk and 5Tk (site "Lira") holes. Values for initial excess pressure observed at opening of many other cavities varied between 0.02 MPa (0.2 atm) to 0.48 MPa (4.9 atm). Information on temperatures in cavities at the time of the opening was obtained by direct temperature measurement in cavities. At least in nine cavities the temperature exceeded 100°C.

TABLE I. Composition of gas effluents from cavities at "Lira" site

Cavity	Gases composition (%) (volumetric)						
	CO_2	CO	CH_4	O_2	N_2	H_2	Heavy hydrocarbons
1-TK	91	≤0.05	2.34	≤0.025	5.6	0.048	
2-TK	8.0	≤0.1	21.3	3.8	27 ÷ 73	0.5 ÷ 3.6	
3-TK	32.4	≤0.05	1.8	≤0.01	51.5	0.19	0.4
4-TK	15.7	≤0.05	3.8	≤0.01	75.7	0.12	1.0
6-TK	17.5	≤0.05	8.0	≤0.01	66.6	2.78	2.1

Scientists from V.G. Khlopin Radium Institute showed that there were at least three different types of gaseous atmosphere in the underground cavities of "Lira" site: (1) CO_2 (1-TK); (2) CO_2-N_2 with methane CH_4 impurity (3-TK, 4-TK, 6-TK; about the same gas composition from T-14 cavity of the "Vega" site); and (3) methanonitric with hydrocarbon dioxide impurities (2-TK and T-15 cavity of the "Vega" site).

Gaseous-liquid inclusions and lenses in a zone of opened fractures could also be the sources of nitrogen and methane. The major part of methane could be formed in decomposition process of organic substances containing rock salt anhydrates layers. The chemical composition of organic compounds in vapour-gaseous mixture of "Lira" site cavities is shown in Tables II and III.

TABLE II. Composition of hydrocarbon gases from some cavities at the "Lira" site (%) (vol)

Gas	Cavities		
	3-TK	4-TK	6-TK
CH_4	1.1 ÷ 2.51	3.5 ÷ 4.2	8.1
C_2H_6	0.23 ÷ 0.37	0.63 ÷ 0.78	1.6 ÷ 1.7
C_3H_8	0.03 ÷ 0.04	0.13 ÷ 0.18	0.21 ÷ 0.27

TABLE III. Activity concentration of Krypton-85 at the "Lira" site

Explosion	^{85}Kr volumetric activity in sample (Bq/m^3·10^5)
1-TK	5.7 ± 0.2
2-TK	5.4 ± 0.1
3-TK	4.0 ± 0.2
4-TK	5.8 ± 0.2

It can be seen that the major hydrocarbon components in these three cavities were: methane –1.1–8.1%, ethane –0.23–1.7%, propane –0.03–0.27%. Beside these compounds, C_3H_6, iC_4H_{10}, nC_4H_{10}, iC_5H_{12}, iC_6H_{14}, nC_6H_{14} were also present. In cavities ready for exploitation ^{85}Kr was practically not present, as it had escaped to atmosphere during the first effluent release.

Tritium, despite multiple gas exchange in the cavity, escaped into the air during repeated effluent treatments (mainly in light-volatile forms of molecular hydrogen and hydrocarbons), but not more than 10% of its total concentration in the cavity. Tritium was mainly left in the cavity and in the fissured zones in the form of HTO oxide. During rapid gas release, when the technological hole has not yet been heated sufficiently, escaped gas does not contain moisture, and tritium is in the gaseous form as HT, HTS and CH_3T. With long-time gas streaming increasing the heat, vapor appeared in the gas, and tritium was predominantly in oxidized form. Tritium was present in different forms; the distribution observed in vapor-gaseous mixture flowing from the E-3 cavity ("Magistral'" site) was: HTO-90%, HT-8%, CH_3T-2%. The results of the investigation of the distribution of tritium chemical forms in vapour-gaseous mixture flowed at "Vega" site are given in Table IV.

Based on results from the investigation on the tritium concentrations and forms in vapour-gaseous mixtures in nuclear explosion cavities in salt rocks, conclusions can be drawn about tritium distribution regularities:

TABLE IV. Tritium distribution in vapour-gaseous mixture (VGM) at opening of the cavities at the "Vega" site

Explosion	Tritium volumetric activity in VGM components, Bq/l					
	HTO	HT	HTS	CH_3T	T-hydrocarbons	ΣT^*
T-3	$7.4 \cdot 10^3$ ÷ $1.8 \cdot 10^6$	$1.1 \cdot 10^5$	$7.4 \cdot 10^4$	$1.1 \cdot 10^6$	$1.5 \cdot 10^5$	$1.5 \cdot 10^6$ ÷ $3.3 \cdot 10^6$
T-6	$7.4 \cdot 10^4$ ÷ $3.3 \cdot 10^7$	$1.5 \cdot 10^5$ ÷ $1.8 \cdot 10^6$	–	$7.4 \cdot 10^5$ ÷ $3 \cdot 10^6$	–	$1.1 \cdot 10^6$ ÷ $3.7 \cdot 10^7$
T-8	$2 \cdot 10^3$	–	$2.5 \cdot 10^4$	–	–	$2.2 \cdot 10^5$
T-14	$2.2 \cdot 10^3$	$1.85 \cdot 10^6$	$3.7 \cdot 10^4$	$2.2 \cdot 10^7$		$2.6 \cdot 10^7$

* Changes in the VGM moisture in release processes resulted in a decrease in the tritium activity concentrations. Upper and lower limits are given.

The original bulk of tritium in flowing VGM is present in a hydrocarbon form, followed by hydrogen, hydrogen sulphide and water in accordance with order of magnitude; Later, increase in the tritium concentration in VGM is observed during flowing process, basically at the expense of moisture increase in VGM, thus increasing the fraction of oxidized tritium form in the total tritium activity concentration in VGM;

The fraction of tritium in oxidized form does not significantly appear at the cavity surface and this fraction is left in the nuclear explosion cavity. This is supported by the fact, that in the VGM release process, the cavity hole plays the part of a "reversed cooler" (it does not manage to heat), and the water vapours condense and flow downwards in the cavity.

Up to now a lot of experimental data are accumulated resulting in information on production contamination stored in nuclear explosions cavities. On the basis of investigations on radionuclides distribution in systems such as salt-gaseous condensate, brine-gaseous condensate, water – methanolic brine – gaseous condensate, possible radionuclides concentrations in gaseous condensate of Orenburg gas-condensate deposit have been estimated. After more than 15 years, which had passed from the time when the first reservoir was created by exploitation, stored gaseous condensate samples have been analyzed with respect to the radionuclides concentrations.

It is evident that long-term storage of gaseous-condensates in reservoirs in contact with rock and brine containing radionuclides does not result in contamination higher than the permissible levels. However, in case of petroleum filling in underground reservoir situation becomes significantly different. Investigation on the radionuclides distribution in systems such as brine-methanol gaseous condensate – petroleum (under different conditions and phase ratios), the radionuclides concentration in organic phase can exceed the permissible levels.

As a summary, the outflow into the atmosphere of ^{85}Kr and tritium from experimental and pilot nuclear explosions with total yield 0.6 Mt, in the period after opening were: $2 \cdot 10^4$ Bq ($5 \cdot 10^3$ Ci of ^{85}Kr) and $\sim 1.3 \cdot 10^{15}$ Bq ($3.5 \cdot 10^4$ Ci of T). The most critical factor influencing the consequences of underground nuclear explosion is the migration of radionuclides into rock massif, and, more exactly into groundwaters (Table V).

TABLE V. Expected and actual contamination levels of gaseous condensate containing radionuclides

		Volumetric activity (Ci/l) Tritium		^{137}Cs	^{125}Sb	^{106}Ru	^{90}Sr
		Light fraction	Hard fraction				
1	Forecast	$n \cdot 10^{-6}$		$N \cdot 10^{-10}$	$n \cdot 10^{-9}$	$n \cdot 10^{-9}$	$n \cdot 10^{-10}$
2	E–1÷E–3	$(1–3) \, 10^{-6}$		$<1 \cdot 10^{-10}$	$(1–10) \cdot 10^{-9}$	$(1–10) \cdot 10^{-9}$	$<1 \cdot 10^{-12}$
3	1TK	$8 \cdot 10^{-8}$	$2.2 \cdot 10^{-7}$	$1 \cdot 10^{-10}$	$<1 \cdot 10^{-8}$	$1 \cdot 10^{-8}$	$<1 \cdot 10^{-11}$
4	2TK	$5.4 \cdot 10^{-7}$		$2 \cdot 10^{-10}$	– " –	– " –	– " –
5	3TK	$5.4 \cdot 10^{-8}$	$5.4 \cdot 10^{-7}$	$1 \cdot 10^{-10}$	– " –	– " –	– " –
6	4TK	$\sim 10^{-6}$		$1 \cdot 10^{-6}$ *	$6 \cdot 10^{-8}$	–	–
	ПДК	$5.4 \cdot 10^{-4}$		$8.1 \cdot 10^{-6}$	$8.1 \cdot 10^{-5}$	$1.1 \cdot 10^{-5}$	$1.5 \cdot 10^{-7}$

*Water was present in sample as emulsion

Observations of radionuclides in groundwater at the "Galit" site for some years have shown that the level at the A-III explosion site was practically not changed (or the change was insignificant). This may indicate that radionuclides coming into groundwaters at the "Galit site", the area of technological sites except for the technological site A-I, occurred as a result of the gaseous precursors to cesium-137 (^{137}Xe) and strontium-90 (^{90}Kr) diffusion at the explosion site, indicating isolation of this water-bearing structure and slow movement rate of groundwaters. In groundwaters at the A-V sites, significant ^{131}Cs concentration was found in the groundwater sampled from the observation hole A-V-2, situated at the distance of 100 m from the main technological hole. Specific activity of ^{137}Cs in 1990 was $4.8 \cdot 10^{-10}$ Ci/l, that is, 30 times lower than PC$_B$. The ^{90}Sr concentration in this water reached values up to $1 \cdot 10^{-10}$ Ci/l, that is, only four times lower than PC$_B$ for water. The tritium activity as HTO in all investigated samples from groundwater at the "Galit" sites was lower than $1 \cdot 10^4$ Bq/l, which was ten times lower then PC$_B$ for drinking water.

Thus, on the basis of the investigations of the groundwaters activity at the "Galit" site in can be seen that ^{137}Cs and ^{90}Sr activity concentrations in observation holes around the cavities of A-II, A-III, A-IV, A-V explosions were lower than PC$_B$ for drinking water, but somewhat higher than the contamination level of surface waters by global fallouts from atmospheric nuclear weapons tests. The exception is the cavity from the A-I explosion. Small oreols of radionuclides in groundwaters, 10 years after the explosion, demonstrate that the migration of radionuclides is still limited.

References

1. Teller E. et al. The constructive uses of nuclear explosives. McGraw-Hill, 1968.
2. Holzer F. Calculation of seismic source mechanisms. Proc. Roy. Soc., 1966, v. 290, N. 1422, pp. 408–429
3. Rawson D., Randolph P. et al. Postexplosion environment resulting from the Salmon event. J. Geoph. Res., 1966, v. 71, N. 14, pp. 3507–3521.

4. Adamsky V.B., Adymov Z.I., Akhmetov E.Z. et al. Peaceful Nuclear Explosions at Bol'shoy Azgir salt dome deposit. Annual Report of the Institute of Nuclear Physics National Nuclear Centre of Kazakhstan Republic, VNIPIPromtechnology, Almaty, 1998 (in Russian).
5. Butkovich T.R. The gas equation of state for natural materials. Lawrence Radiation Laboratory, University of California, Livermore, CA, UCRL-14729, 1967.
6. Rawson D., Boardman C., Jaffe-Chazan N. The environment created by a nuclear explosion in salt. Project Gnome, report PNE-107F, Livermore, CA, 1964.
7. Kormer S.B., Sinitsyn M.B., Kirillov G.A., Urlin V.D. Experimental determination the temperatures of shock compressed NaCl and KCl, and their curves melting up tp 700 kbar. Journal of Experimental and Theoretical Physics Letters (JETP Letters), 1965 (in Russian).
8. Kahn J.S., Smith D.K. Mineralogical investigations in the debris of the Gnome event near Carlsbad, New Mexico. Amer. Mineral., 1966, v. 51, pp. 1192–1199.
9. Griffiths P.R., Brown C.W., Lippincott E.R., Dayhoff, M.O. Termodynamic models in cosmochemical systems. Geochim. Cosmochim. Acta, 1972, v. 36, N. 2, pp. 109–129.
10. Yang J. Decomposition of solid substances kinetics, Moscow, Mir, 1969 (in Russian).
11. Edwards A.L., Holzman R.A. Thermal effects of nuclear explosion in salt. The Salmon experiment. Naturwissen, 1968, Bd. 55, N. 1, s. 18–22.
12. Rawson D.E., Taylor R.W., Springer D.L. Review of the Salmon experiment a nuclear explosion in salt. Nuturwissen, 1967, Bd. 54, N. 20, s. 525.
13. Adamsky V.B. Growth cavity by explosion and accompanying seismic effect. Issues of Atomic science and technic. Theor. Appl. Phys., issue 4, pp. 8–17, TsNIIAtomInform, M., 1991
14. Geck F.W. Zs. Naturforsch, 1957, Bd. 12a, s. 562.
15. Mott G., Gerni R. Electronic process in ionic crystals, M., 1950 (in Russian).

CHAPTER 4

RADIATION AND HYDRO-CHEMICAL INVESTIGATION AND MONITORING OF TRANSBOUNDARY RIVERS OF KAZAKHSTAN

K.K. KADYRZHANOV[1], H.D. PASSELL[2],
V.P. SOLODUKHIN[1], S. KHAZHEKBER[1],
V.L. POZNYAK[1], E.E. CHERNYKH[1]
[1]*Institute of Nuclear Physics NNC RK, Ibragimov str. 1, 050032, Almaty, Kazakhstan Republic*
[2]*Sandia National Laboratories, P.O. Box 5800, Albuquerque, NM87185-1373, USA*

Abstract: The present paper discusses the general state of the aqueous resources of Kazakhstan and provides an overview of the factors influencing the state of the environment in the catchment areas of the four main transboundary rivers in Kazakhstan (the Ural, the Syrdariya, the Ile and the Irtysh rivers). In addition, it is proposed to put up a programme aiming to monitor the contamination of these river systems using nuclear analytical methods. In view of the activity arrangements, experience gained from the implementation of the international project "Navruz" is assumed to be used. The "Navruz" project was devoted to ecological survey and creation of the monitoring system for main rivers of the Central Asia – the Syrdariya River and the Amudariya River – over the lands of Uzbekistan, Kyrgyzstan, Tajikistan and Kazakhstan.

Keywords: radionuclides, metals, monitoring, transboundary rivers, Kazakhstan

1. Introduction

Water resources in the Republic of Kazakhstan are limited and they suffer from a relatively high degree of pollution and desiccation. The disintegration of the Former Soviet Union has made matters worse due to reduced efforts to coordinate the regulation of river flows and water quality monitoring. The lack of reliable information on the state of the of river environment and potential sources of contamination represent problems related to both environmental risk and socio-psychological stress. Thus, it is of importance to the Republic of Kazakhstan to develop and implement a monitoring programme with respect to radionuclide and heavy metal contamination.

2. Potential Sources Contributing to River Contamination

Some general information on the largest transboundary rivers in Kazakhstan are given in Table I.

TABLE I. General information on the main trans-boundary rivers of Kazakhstan

River name	Trans-boundary countries	Total river length (km)	Water flow into Kazakhstan (m^3/sec)
Ural	Russia	2,428	400
Syrdariya + Naryn	Uzbekistan	3,019	450
Ile	China	1,001	330
Irtysh + Black Irtysh	China, Russia	4,248	800

All these rivers are of great importance for the population of Kazakhstan. However, the presence of a number of large industrial centres, agricultural enterprises and potential sources of radioactive contamination within the catchments of these rivers cause concern with respect to chemical contamination including radionuclides. An overview of facilities representing potential environmental risks is given in Fig. 1 [1].

The main factors influencing the state of the environment in the catchment areas of the four main transboundary rivers in Kazakhstan; the Ural, the Syrdariya, the Ile and the Irtysh rivers, are briefly discussed in the following:

- The Ural springs from the southern spurs of the Ural mountain range, an area in which mineral resource industry and metal manufacture are widely developed. Within Russia, potential sources of radioactive contamination in the Ural and its tributaries include radioactive fallout from the atmospheric nuclear explosion (40 kt) during the Totski troop exercises (14 September 1954) and radionuclide releases from the Mayak PA enterprise due to the Kyshtym accident (1957) and Karachaev event (1967; wind transport of dry radioactive sediments from Lake Karachai). Within Kazakhstan, the Lira sites (six underground nuclear explosions), the Kapustin Yar test site and a number of industrial enterprises devoted to mining and processing of organic and mineral resources are potential sources of contamination.
- The Syrdariya River is the main freshwater resource of South Kazakhstan. Potential sources of contamination within the river catchment include numerous industrial sites (mining, metallurgy, chemical industry etc.) in neighbouring countries. These include the mining complex Mailuu-Su (Kyrgyzstan); the uranium processing enterprise Vostokredmet in Chkalovsk City (Tajikistan), numerous industrial enterprises and plants in Uzbekistan (Kibrai, Angren, Yangiabad, Navoi and other cities). In addition, these neighbouring countries (most notably, Uzbekistan) carry out an intensive agricultural activity within the Syrdarya River catchment. Kazakhstan also has numerous industrial enterprises on uranium mining and processing (Shyili town) and heavy metals (Shymkent City) within the catchment of this river.

- The sources of the Ile and Irtysh rivers are in China, being a country of high population density. China is carrying out an intensive agricultural and industrial activity within the catchment areas of these rivers. It is also hypothesized that, radioactive fallout from nuclear explosions performed at the Lop Nor test site (more than 43 explosions) could be a potential source of radioactive contamination of these rivers. Of particular concern are the 16 air and ground explosions performed during 1964–1980 (W = 10 Mt). There are also a large number of Kazakh agricultural enterprises and mining-and-metallurgical plants within the catchment areas of these rivers. It is significant that runoff from the Semipalatinsk test site, on the territory of which 456 nuclear explosions were performed, may contribute to the radionuclide contamination of the nearby Irtysh River.

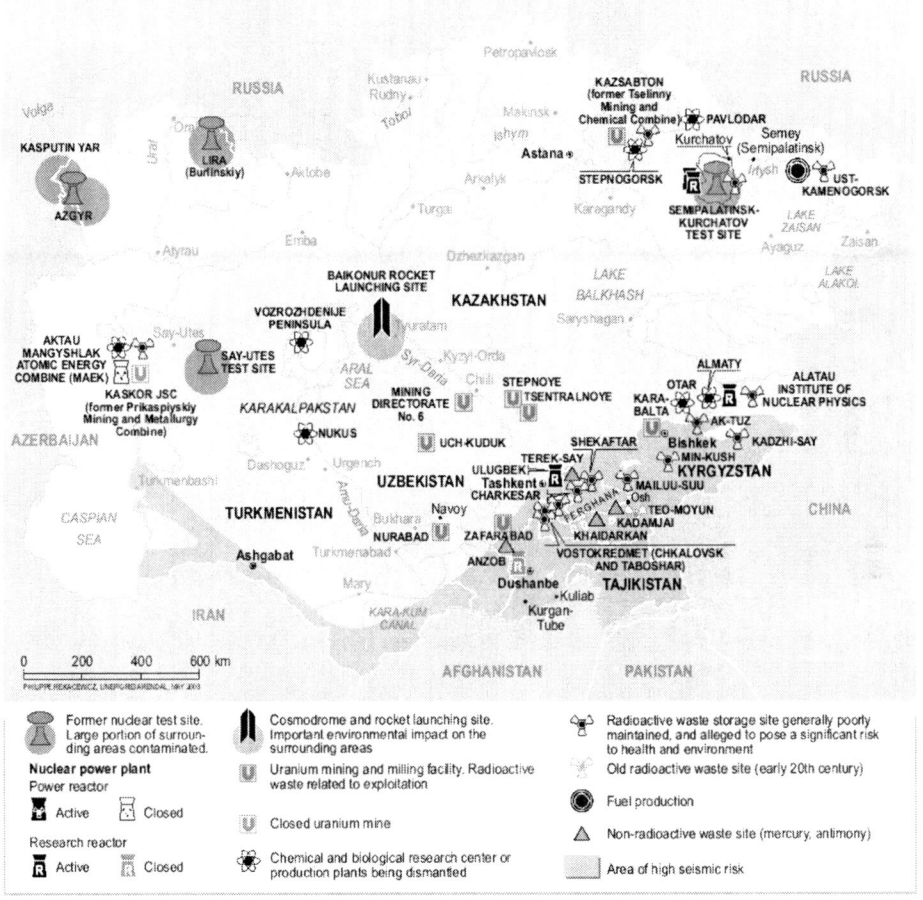

Figure 1. Scheme of arrangement of ecologically dangerous facilities

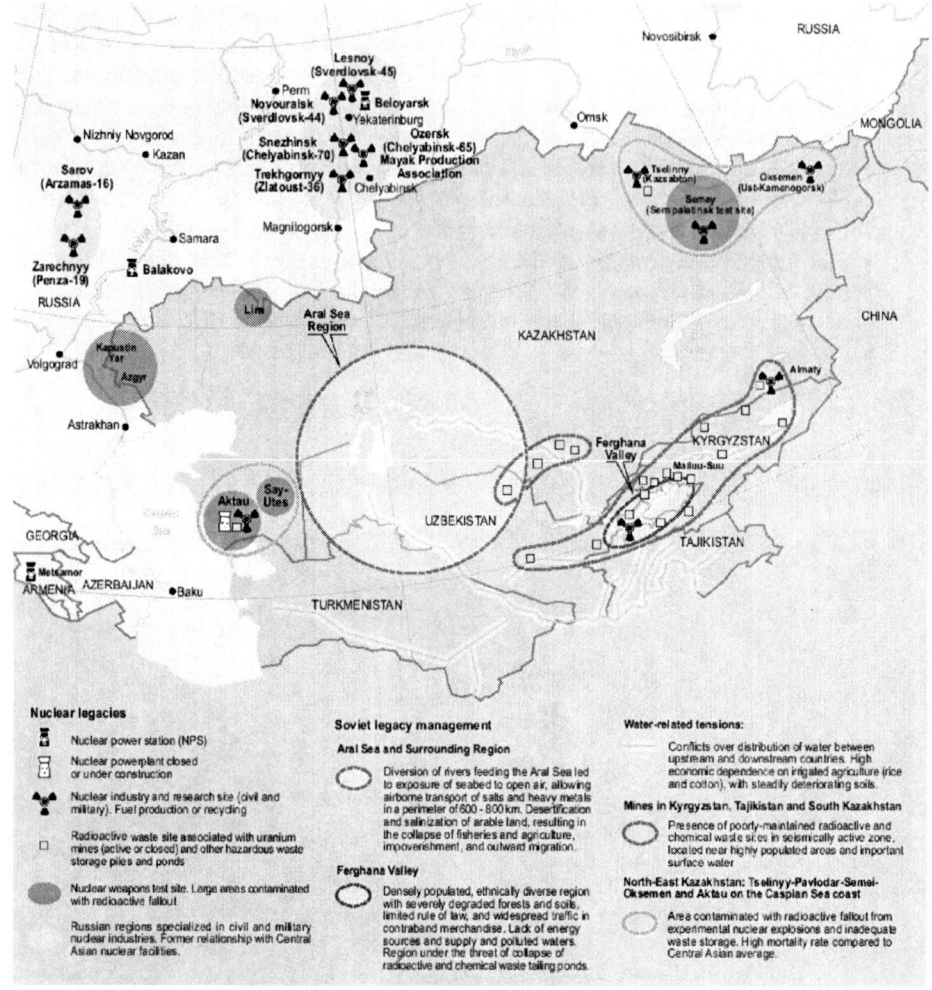

Figure 2. Scheme of main problems of the Central Asia in the area of environment and safety

The current situation is of special concern for the world community and international organizations (UNEP, UNDP, OSCE, NATO, etc.). Figure 2 illustrates the main problems related to environmental security in Central Asia [1].

In order to assess the probabilities and consequences of events that may influence the water quality of the rivers in question, reliable information is needed. A lot of information may be obtained through monitoring of the transboundary rivers by means of modern nuclear analytical methods of analysis. It is very important that all the countries concerned participate in the proposed monitoring programme. Organizations with nuclear competence from Uzbekistan, Kyrgyzstan, Tajikistan and Kazakhstan in collaboration with Sandia National Laboratories, USA, have since 2000 been involved in monitoring of the large Syrdarya and Amudarya rivers within the framework of the Navruz project [2–4]. Experience from this project was used in the preparation of the International project "Caspian Rivers" [5], which deal with the planning of a monitoring programme for Volga, Kura, Ural and Emba rivers. These rivers represent 90% of all the discharges into the Caspian Sea. Participants of this project are scientific institutions from Russia, Georgia, Armenia, Azerbaijan and Kazakhstan as well as Sandia National Laboratories. A similar international project on radiation and hydro-chemical monitorring of the Ile and Irtysh rivers with participation of scientific institutions from China, Kazakhstan and Russia has been discussed.

3. Conclusions

Environmental concerns related to contamination of transboundary rivers in Kazakhstan from contaminants including radionuclides are increasing. All countries concerned need to take part in the environmental management of these freshwater resources, including a monitoring programme. The nuclear analytical techniques developed in the framework of the International Project "Navruz" can be useful in the implementation of such a programme.

References

1. Martino, L., Carisson, A., Rampolla, G., Kadyrzhanova, I., Svedberg, P., Denisov, N., Novikov, V., Rekacewicz, Ph., Simonett, O., Skaalvik, J.F., Pietro, D., Rizzolio D. and Palosaari, M. (2005) Environment and Security. *Transformation Risks into Cooperation. UNEP, UNDP, OSCE, NATO.* ISBN: 82-7701-035-4.
2. Yuldashev, B.S., Salikhbaev, U.S., Kist, A.A., Radyuk, R.I., Barber, D.S., Passell, H.D., Betsill, J.D., Matthews, R., Vdovina, E.D., Zhuk, L.I., Solodukhin, V.P., Poznyak, V.L., Vasiliev, I.A., Alekhina, V.M. and Djuraev, A.A. (2005) Radioecological monitoring of transboundary rivers of the Central Asian region. Radionuclide contamination in the Syr Darya river basin of Kazakhstan; Results of the Navruz Project. *Journal of Radioanalytical and Nuclear Chemistry*, **263**(1): 219–228.
3. Kadyrzhanov, K.K., Barber, D.S., Solodukhin, V.P., Poznyak, V.L., Kazachevskiy, I.V., Knyazev, B.B., Lukashenko, S.N., Khazhekber, S., Betsill, J.D. and Passell, H.D. (2005) Radionuclide contamination in the Syr Darya river basin of Kazakhstan; Results of the Navruz Project. *Journal of Radioanalytical and Nuclear Chemistry*, **263**(1): 197–205.
4. Solodukhin, V.P., Poznyak, V.L., Kazachevskyi, I.V., Knyazev, B.B., Lukashenko, S.N. and Khazhekber, S. (2004) Some peculiarities of the contamination with radionuclides and toxic elements of the Syrdarya river basin, Kazakhstan. *Journal of Radioanalytical and Nuclear Chemistry*, **259**(2): 245–250.
5. Passell, H.D., Barber, D.S., Kadyrzhanov, K.K., Solodukhin, V.P., Chernykh, E.E., Arutyunyan, R.V., Valyaev, A.N., Kadik, A.A., Stepanetts, O.V., Vernadsky, V.I., Alizade, A.A., Guliev, I.S., Mamedov, R.F., Nadareishvili, K.S., Chkhartishvili, A.G., Tsitskishvili, M.S., Chubaryan, E.V., Gevorgyan, R.G. and Puskyulyan, K.I. (2005) The international project of radiation and hydrochemical investigation and monitoring of general Caspian rivers. *5th International Conference "Nuclear and Radiation Physics"*, Vol 1, Almaty, Kazakhstan, 2006, 69–76.

CHAPTER 5

TRITIUM IN STREAMS, WELL WATERS AND ATOMIC LAKES AT THE SEMI-PALATINSK NUCLEAR TEST SITE: PRESENT STATUS AND FUTURE PERSPECTIVES

P.I. MITCHELL[1], L. LEÓN VINTRÓ[1],
J.G. HOWLETT[1], M. BURKITBAYEV[2],
N.D. PRIEST[3], YU.G. STRILCHUK[4]
[1]*UCD School of Physics, University College Dublin, Belfield, Dublin 4, Ireland*
[2]*Department of Inorganic Chemistry, Al-Farabi Kazakh National University, Almaty, Kazakhstan Republic*
[3]*School of Health and Social Sciences, Middlesex University, Queensway, Enfield EN3 4SA, UK*
[4]*Institute of Radiation Safety and Ecology, National Nuclear Centre, Kurchatov, Kazakhstan*

Abstract: The database on tritium concentrations in streams, atomic lakes, ground waters and well waters at the Semipalatinsk Nuclear Test Site (STS) is by no means extensive, given the scale of nuclear testing undertaken at the site in the past. Here, we highlight the extent of the present database by summarising all of the readily accessible (published) data on tritium levels in the various waters of interest. In addition, we present new data on tritium gathered in the course of fieldwork in 2004 and 2005, by participants in the SEMIRAD 2 Project. We also review the status of present knowledge in regard to the mobility of tritium within and off the STS, by focussing on specific case studies, and offer suggestions as to the shape future research endeavours might take that, in our opinion, would be most cost effective.

Keywords: tritium, streams and lakes, ground water, Semipalatinsk test site

1. Introduction

Tritium is a naturally occurring radionuclide produced mainly from interactions between cosmic-ray neutrons and nitrogen in the upper atmosphere, via the reaction $^{14}N(n,T)^{12}C$.

It is also an important radioactive component of liquid and gaseous discharges from nuclear power plants and spent nuclear fuel re-processing installations, although the yield from nuclear fission (and, by extension, atomic weapons) is actually quite low, being on the order of 10^{-2}%. In contrast, the yield from nuclear fusion (i.e. thermonuclear weapons) is some three to four orders of magnitude higher on a pro rata yield (TNT equivalent) basis and is closely related to the design of the nuclear device (e.g. fission–fusion–fission type) and the environment within which it is detonated [1].

It is well established that tritium can displace hydrogen in water, to form tritiated water (HTO/TTO) and migrate with ground and surface waters. Hoffmann et al. [2] have shown that in excess of 99.9% of the tritium liberated by an underground nuclear explosion at the Nevada Test Site (US) exists as HTO. Others have suggested a figure of ~98% as HTO, with ~2% as HT [3]. Regardless, essentially all of the released tritium is assumed to form water, either by oxidation or exchange.

It has been estimated that the tritium concentration in the crushed zone produced by a contained underground test involving a 1 Mt fusion device triggered by a 10 kt fission device would be about 100 MBq dm^{-3}, assuming that the exchange of tritium between the tritiated water and the rock matrix is negligible [4, 5]. The actual concentrations attained would, of course, be dependent on factors such as the extent of the crushed zone, the amount of venting of tritium that occurred, the scale of reactions between tritium and minerals of the formation, and the water content of the formation. Outside of this zone the concentration of tritium would, obviously, be reduced by dilution and dispersion in nearby groundwater. Interestingly, concentrations of tritium in waters flowing from tunnel portals and adits in the Degelen Massif on the Semipalatinsk Nuclear Test Site (STS) in the period 1996–2000, inclusive, were reported to be in the range 0.02–1.7 MBq dm^{-3} [6, 7], which would appear to be in broad accord with the above-mentioned estimate of 100 MBq dm^{-3} for water in the crushed zone, when it is appre-ciated that the devices exploded within the Degelen tunnels had all relatively low yields (≤50 kt) by comparison [8].

Tritium is regarded as a key nuclide for the study of groundwater migration and for monitoring the contamination of groundwater by radionuclides from underground nuclear tests. In fact, it is considered a 'leader' nuclide in the monitoring of both radioactive pollution and organic contamination in groundwater, as the migration rate of tritium is the same as the flow rate of groundwater. For the record, the rate of flow, F_a, of a given radionuclide with groundwater can be expressed, to first order, by a simple equation [9] of the form

$$F_a = \frac{F_w}{1 + \rho K_d} \qquad (1)$$

where F_w is the flow velocity of the water, ρ is the ratio of the mineral weight to that of water per unit volume for the given mineral (usually $\rho = 4$–5), and K_d is the distribution coefficient for the mineral (i.e. the ratio of the number of ions bound in the solid phase to the number in aqueous solution).

One can appreciate from this formula that the flow rate of a given radionuclide with groundwater will be hundreds or thousands of times smaller than that of the water, given that K_d values for typical fission products are in the range 10^1–10^5. Even under large groundwater flow velocities (say 10 m per day), the migration, for example, of ^{90}Sr over a distance of 1 km is predicted to take at least a hundred years (Table I).

TABLE I. Mobility of selected radionuclides in groundwater (a groundwater flow velocity of 10 m day^{-1} has been assumed for the purpose of the calculation)

Radionuclide	$T_{1/2}$ (year)	K_d	F_a (my^{-1})
^{3}H	1.2 10^1	0	3,650
^{36}Cl	3.0 10^5	1	730
^{99}Tc	2.1 10^5	100	9
^{129}I	1.6 10^7	300	3
^{90}Sr	2.9 10^1	1,000	0.9
^{239}Pu	2.4 10^4	100,000	0.009

Needless to add, in the case of tritium, the Kd is taken to be zero. Underground testing is known to have resulted in severe damage to the rock structure in the Degelen Massif and the Balapan test field, creating numerous fissures, rockslides, chimney collapses and surface subsidence. It follows that actual migration rates are expected to be considerably higher than predicted solely on the basis of the above formula, which is applicable only to groundwater flow through porous or semi-porous geologic media.

Since the first nuclear test by the former Soviet Union in August 1949, a total of 456 such tests were conducted at the STS in the period 1949–1989. Approximately 340 of these were underground tests, most of which were carried out following the signing of the Limited Test Ban Treaty (the Moscow agreement) in 1963 [10, 11], mainly within horizontal tunnels on the Degelen Massif (~215) and in vertical shafts on the Balapan test field (~107). Both of these areas are situated in the East Kazakhstan Oblast on the STS to the west and north, respectively of the settlement of Sarzhal (pop. ca. 2,000), which lies just outside the test site boundary (Fig. 1).

Following a preliminary assessment of the radiological situation at the STS in 1998 by experts from the International Atomic Energy Agency [12], the Agency urged, amongst other matters, that studies be conducted to investigate the radiological status of local sources of drinking water, in order to determine whether these waters had been contaminated with radioactivity from the many tests that had been conducted, particularly those carried out underground.

Here, we attempt to summarise the existing database on tritium concentrations in streams, atomic lakes, groundwaters and well waters of the STS. Included are data (much of it new) gathered in the course of fieldwork in 2002, 2004 and 2005 by participants in the NATO SfP-funded SEMIRAD 1 [13] and SEMIRAD 2 projects (Fig. 1). Some of the latter data have been determined using electrolytic enrichment.

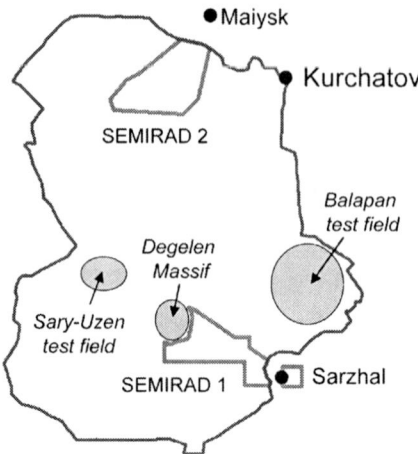

Figure 1. Outline of the STS boundary with the SEMIRAD 1 and SEMIRAD 2 study areas indicated

In addition, we review the status of present knowledge in relation to the mobility of tritium within and off the STS, by focussing on specific case studies. Finally, we offer suggestions as to the shape future research endeavours might take that would likely be most cost effective in the longer term.

2. Methodology

2.1. DATABASE OF PUBLISHED TRITIUM LEVELS IN NATURAL WATERS (AND SOILS) OF THE STS

We have carried out an extensive search for articles and reports on tritium concentrations (and mobility) in streams, atomic lakes, ground waters, well waters and soil waters at the STS published in the past 25 years, utilising a number of well-known databases, including Sci Central, Ingenta Connect, British Library Direct, Inist, Scirus and ISI Web of Knowledge. Surprisingly, only a few publications of relevance were identified by these means. On the other hand, publications in the form of annual research bulletins by the NNC RK and its research institutes (i.e. INP and IRSE), as well as published conference and workshop proceedings (mainly in Russian), yielded a considerable, though by no means extensive, body of valuable data.

For convenience, we have grouped the data discovered under five main headings, namely, (i) tritium in water flushes from test tunnel portals in the Degelen Massif; (ii) tritium in atomic lakes within the STS; (iii) tritium in groundwater following underground cratering explosions; (iv) tritium in natural streams, lakes (including salt-pans) and rivers of the STS; and (v) tritium in domestic/farm wells sited on or in close proximity to the STS.

2.2. FIELD SAMPLING

Sampling surface waters of the STS for the purpose of determining their radionuclide content is straightforward, requiring only that precautions be taken to ensure that the samples are not contaminated in the course of sampling by, for example, contact with soil. Collecting groundwater samples is considerably more difficult, particularly if old observation boreholes (artesian wells) have to be cleaned out before cylindrical samplers are deployed as, for example, was the case in the ISTC-funded project (#K-810), which examined the mobility of various radionuclides (including tritium) in groundwater near the epicentre of an underground cratering explosion conducted at Sary-Uzen in October 1965 [14]. Full details of the sampling strategies employed in the various studies to which we refer in this review may be found in published reports and articles on these studies.

For our part, we have sampled streams, well waters, rivers, atomic lakes and seasonal lakes in the course of field campaigns (SEMIRAD 1 and SEMIRAD 2) to the Degelen/Tel'kem/Sarzhal, Balapan/Chagan River and Maiysk regions of the STS in July of 2002, 2004 and 2005, respectively. Specifically, a volume of approximately 1 l of water was collected at each sampling site and sealed in a clean air-tight bottle after the bottle had been rinsed thoroughly with a similar volume of sample water for the purpose of pre-equilibration.

2.3. TRITIUM ANALYSIS BY DIRECT MEASUREMENT

Upon return to the laboratory, the samples were filtered through 0.45 μm cellulose nitrate membranes and a 5 ml aliquot of each was added to high quality 30-ml polyethylene counting vials pre-loaded with 10 ml aliquots of water-miscible liquid scintillant (Ultima Gold AB supplied by Packard) and shaken vigorously for a few minutes. Counting was carried out using an LKB-supplied Wallac Quantulus low background liquid scintillation counter, the settings of which had been optimised for low-level tritium measurement.

Counting efficiency, at $0.294 \pm 0.004(2\sigma)$ for a window setting of 20–250 channels and a pulse shape analysis (PSA) level of 117, was determined by gravimetric dilution of a certified ^{3}H-labelled water standard (product code: *TRY 44*) supplied by Amersham International plc. (UK). The mean ($n = 9$) laboratory blank count rate was determined to be $1.04 \pm 0.04(2\sigma)$ min^{-1}. The counting time for all but the most active samples was set at 1–20 h and the corresponding minimum detectable activity concentration ranged from 9 to 1.7 Bq(^{3}H) dm^{-3} on the basis of Currie's criterion [15].

2.4. TRITIUM ANALYSIS FOLLOWING ELECTROLYTIC ENRICHMENT

A small sub-set (n = 7) of the samples from the SEMIRAD 2 campaigns of 2004 and 2005, all of which were found to have very low tritium concentrations upon measurement by the direct method, were electrolytically enriched at one of our laboratories (UCD) after the method of Baeza et al. [16] and measured by liquid scintillation counting, as described above. The activity concentration was then calculated using the approach of Taylor [17]. Good agreement between the results as determined by the direct method and following enrichment is apparent from an examination of the data given in Table II below. The minimum detectable activity concentration, following enrichment, proved to be an impressive 0.20 Bq(^{3}H) dm^{-3} for a counting time of 20 h.

TABLE II. Comparison of tritium concentrations (±2σ) in selected water samples from The STS measured by low-background LSC, with and without electrolytic enrichment. Note: one tritium unit (1 T.U.) = 1 atom of tritium per 1,018 atoms of hydrogen and is equivalent to approximately 0.1182 Bq(^{3}H) dm^{-3}(H$_2$O).

Source of sample/year sampled	Enrichment (Bq dm^{-3})	Direct LSC (Bq dm^{-3})
Irtysh river/2004	2.3 ± 0.3	<9
Irtysh river/2005	2.5 ± 0.3	<6
Chagan river near confluence with Irtysh/2005	6.5 ± 0.6	6 ± 3
Sarzhal well/2004	3.9 ± 0.4	<9
Ashyozek stream (SEMIRAD 2 study area)/2005	2.6 ± 0.2	<6
Ashyozek stream (SEMIRAD 2 study area)/2005	1.3 ± 0.1	<6
Almaty lake (Tien Shan mountains)/2004	1.9 ± 0.3	<9

3. Status of the Tritium Database (as Published)

Given the great scale of nuclear weapon and device testing undertaken in the past, published data on tritium concentrations in streams, atomic lakes, ground waters and well waters at the STS are sparse, by any standards. Significant gaps are apparent in the database; in particular, the monitoring of dome
has been intermittent to say the least. With the important exception of water flushes flowing from test tunnels and adits in the Degelen Massif, time series data on tritium concentrations are rare. Moreover, there appear to be few if any published data on tritium concentrations in soil waters of the STS, nor any on concentrations of organically bound tritium in biota.

This situation is of considerable concern as tritium is correctly considered to be a leader nuclide in hydrological terms for other water-borne nuclides and any abrupt increase in its manifestation in environmental waters should trigger regular and comprehensive monitoring of the zone or catchment involved. This said, we recognise that considerable efforts have been made by Kazakh state research institutes in recent years to study tritium levels at the STS and that some of the resulting data may have been published in internal reports or reports with limited circulation, not easily accessible by us. We are also aware that studies to monitor the mobility of tritium at the STS have recently been given high priority by the same authorities, and this is to be commended.

In the past, during the active period, when the STS was under Soviet control, extensive studies of groundwater contamination were, we understand, undertaken; however, few of the resulting data would appear to be accessible today. An obvious exception is data on the Sary-Uzen cratering explosion of 1965, which were published by Izrael et al. [18] as early as 1970.

4. Case Studies

4.1. TRITIUM IN WATER FLUSHES FROM TEST TUNNEL PORTALS IN THE DEGELEN MASSIF

Various studies over the past decade have shown that radioactivity continues to be flushed out from test tunnel portals (adits) with water manifestations in the Degelen Massif, even after their closure. This radioactivity largely accumulates in permanent and temporary streambeds on the STS, though the potential for more distant migration, particularly of the more mobile species (e.g., ^{3}H, ^{14}C, ^{36}Cl, ^{99}Tc), is an obvious concern.

Annual mean tritium concentrations in streams flowing from some 14 test tunnels in the 10-year period 1996–2005, inclusive are given in Table 3. Concentrations vary from some tens to some few thousands of kBq dm^{-3}, with tunnel #177 consistently exhibiting the highest manifestation. It also shows the highest ^{137}Cs concentrations in the same period [22]. In the case of individual flows, concentrations show considerable variations over time, depending on season and the level of precipitation (and snow melt) which preceded sampling. Nevertheless, in almost all cases, the overall trend is one of reducing tritium concentrations, consistent with the relatively short half-life of tritium and on-going dilution of water within (and about) the crushed zone by the infiltration of precipitation. The closing of adits by the caving of overlying rock or by plugging with gravel and concrete has significantly reduced water flow and, thus, the washout of radioactivity [8]. Consequently, annual fluxes of tritium in streams from adits that continue to manifest water have fallen in virtually all cases [23]. However, it is almost certainly the case that some of the contaminated water has found alternative exits to the topographic surface of the Massif and thence to natural streams or, perhaps, formed new (seasonal) streams.

Streams flowing radially out to the north, northeast and northwest drain into a closed hydrologic basin, where tritium is lost to the atmosphere by evaporation, while those flowing to the south and southeast would appear to drain into the flood plain of the Irtysh or become lost in the sandy soil.

That the evaporation of tritium is an important loss mechanism on the STS (and other test sites for that matter) is clearly evidenced by recently published data, which show air water-vapour concentrations in the range 90–21,000 Bq dm^{-3} close to Degelen tunnel portals and air concentrations of 1–240 Bq m^{-3} at the same sampling sites [22]. Climatically, temperatures can reach 40°C and higher on the STS in summer and humidity is often relatively low. Snow, of course, is a feature of winter on the territory and tritium concentrations in the range 10–29,000 Bq dm^{-3} have been reported for snow at Degelen [22], which are fully consistent with the air water-vapour concentrations referred to herein.

TABLE III. Mean annual tritium concentrations in kBq dm^{-3} (rounded to two significant figures) recorded in stream waters flowing from test tunnel portals in the Degelen Massif, in the period 1996–2005, inclusive [Sources: 6, 7, 19, 20, 21, 22]

Tunnel code #	1996	1997	1998	1999	2000	2001	2002	2003	2004	2005
A-1	340	250	240	170						
11	190	0	130							
104	280	240	310	230	190	240	220	220	220	250
151	280	(250)								
152	360	(290)	40							
156-T	21	30	47							
165	440	460	430	350	260		230			
176	670	630	640	510	480	530	450	340	360	500
177	1,300	1,700	1,300	1,100	770	600	600	640	680	430
503	190	180	250	170	180	150	130	110	170	140
504	440	520	410	260	260	280	260	260	260	230
506	100	(150)								
511	200	430	320	290	210	190	190	180	160	110
609	42	48	33	16	31	30	30	30	30	

Note: figures in parentheses represent single measurements

4.2. TRITIUM IN ATOMIC LAKES WITHIN THE STS

Published data on tritium concentrations in the waters of the atomic craters at Tel'kem 1 and Tel'kem 2 (Fig. 2) are few, and not in good agreement, year on year. However, it is clear that the levels (ca. several hundred Bq dm^{-3}) are, relatively speaking, low (Table IV)

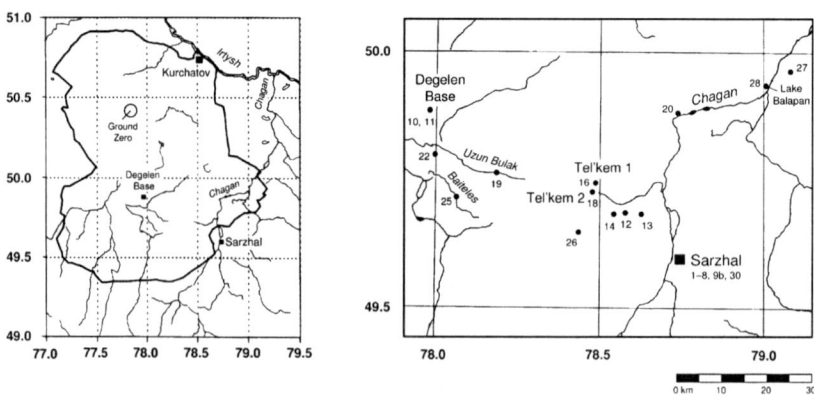

Figure 2. The SEMIRAD 1 study area showing the locations of streams, well waters and atomic lakes sampled in the July 2002 survey

being approximately two orders of magnitude above the tritium background in surface waters of the Northern Hemisphere at the present time (Table V). This is consistent with the very low yields (<1 kt) of each of these explosions, carried out at a depth of about 30 m in sandstone in late 1968 [9].

TABLE IV. Tritium concentrations (Bq dm^{-3}) in water samples taken from atomic lakes on the STS [Sources: 7, 21, 24]

Atomic lake (code #)	2000	2001	2002	2003	2004	2005
Tel'kem 1 (#16)			84	640		
Tel'kem 2 (#18)			178	550		
Lake Balapan (#28)			12,600	15,000		

In contrast, tritium levels in the waters of the large atomic crater that is Balapan Lake, at approximately 15 kBq dm^{-3}, are relatively high compared to the levels recorded in the Tel'kem craters, reflecting the thermonuclear nature of the explosive device used to create this man-made lake adjacent to the Chagan River. The borehole test ("Chagan 1004") involved, had an explosive yield of 100–150 kt and was carried out at a depth of 175 m in aqueous sandstone mixed with coal-clay shale in January 1965 [9].

TABLE V. Tritium in selected waters in the Northern Hemisphere, 1976–present

Water system	Concentration (Bq dm^{-3})	Period	Reference
Yenisei, Siberia	2.4–2.8	2001	[25]
Yenisei & Lena, Siberia	3.7–4.3	1998–1999	[26]
Ebro, Spain	3.3–6.7	1994	[27]
Danube, Serbia	2.4–15.9	1976–1990	[28]
Drinking water, Turnstall, UK	0.9–3.0	1987–1995	[29]
Drinking water, Madrid	2–3	1994–1995	[29]
Tap water, Seville	0.15–0.47	2002–2003	[30]
Almaty Lake, Kazakhstan	1.9	2004	This study
Tap water, Dublin, Ireland	0.9	2006	[31]

4.3. TRITIUM IN GROUNDWATER FOLLOWING UNDERGROUND CRATERING EXPLOSIONS – THE SARY-UZEN EXPERIENCE

Groundwater contamination levels surrounding the epicentral zone of the Sary-Uzen underground cratering explosion (1.1 kt yield) of October 1965 was investigated by Izrael et al. [18] in the months immediately following detonation, using observation boreholes specifically drilled for the purpose prior to the explosion. At this site (Fig. 1), groundwaters basically discharge east and south towards the dry Sary-Uzen riverbed. A year post-explosion, elevated total beta concentrations (data on individual beta-emitting radionuclides were not recorded at the time) remained only east and south of the epicentre, to distances of some few hundred metres, at most [17]. Groundwaters at the site proved to be essentially stagnant (flow velocity estimated to be ~0.13 m d^{-1}). Follow up studies (ISTC Project K-810), 35 years later, indicated tritium concentrations diminishing rapidly with distance along a broad south-easterly 'radial' (Table VI), to about 100–250 Bq dm^{-3} at 730 m [14]. Concentrations along all other radials examined proved to be lower and, in most cases, much lower.

TABLE VI. Tritium concentrations in groundwater sampled in 2003 from observation boreholes in the vicinity of the cratering explosion conducted underground at Sary-Uzen in October 1965 [Source: 14]

Radial from epicentre	Distance (m)	Concentration (Bq dm^{-3})
10°	520	3
55°	360	<5–17
100°	510	202–324
145°	370	1,400–1,700
145°	730	102–255
234°	370	50
279°	320	<5
279°	510	<5
325°	730	<5

4.4. TRITIUM IN NATURAL STREAMS, LAKES (INCLUDING SALT-PANS) AND RIVERS OF THE STS

Tritium concentrations in natural streams draining the Degelen massif, such as Beiteles, Kara-Bulak, Takhtakushuk and Uzun-Bulak, are in the order of 100–200 kBq dm^{-3} at the present time (Table VII). In most cases, measurements were recorded close to the foot of the Massif or within some few kilometres thereof. Some of these streams disappear into the sandy soil after 10 or 15 km or pass underground.

Wide variations have been noted in tritium concentrations in the Chagan River over time, with very high concentrations occasionally recorded to the northeast of Balapan Lake, proximate to the Balapan Test Field (see Sect. 5.1 below for further discussion and recommendations).

On the other hand, in the SEMIRAD 2 study area, lying to the south of the settlement of Maiysk (Fig. 1), tritium concentrations in stream waters were, as expected, very low. A similar observation applies to the Irtysh River, where concentrations in the range 2 to <9 Bq dm^{-3} were recorded in the period 2002–2005 (Table VII), which seemingly evidence little export of tritium from the STS via the Irtysh catchment, at least during the summer season when the samples were collected.

TABLE VII. Tritium concentrations (Bq dm^{-3}) in natural streams and rivers of the STS [Sources: 7, 21, 24, 32]

Stream/river	1999	2000	2001	2002	2003	2004	2005
Uzun-Bulak	140,000	154,000		150,000			
Kara-Bulak	104,000	212,000					
Quarry near Kara-Bulak	200,000	105,000					
Beiteles	140,000	165,000		133,000			
Takhtakushuk				214,000			
Borehole (#385)	222,000	280,000					
Chagan – upper reaches				36			
Chagan – NE of Balapan Lake						196	281,000
Chagan – lower reaches						159	6.5
Ashyozek							2
Irtysh at Kurchatov				9		2.3	2.5

A small number of lakes and salt-pans (which had not fully dried out) were also sampled in the SEMIRAD 2 study area and, again, the concentrations proved to be extremely low (see Tables XI and XII below for the actual values recorded).

4.5. TRITIUM IN DOMESTIC/FARM WELLS SITED ON OR IN CLOSE PROXIMITY TO THE STS

4.5.1. *The SEMIRAD 1 study area*

Tritium concentrations in well waters sampled in the course of the SEMIRAD 1 campaign of July 2002 (Fig. 2) are summarised in Table VIII. The data show that levels of tritium in domestic well waters from within the settlement of Sarzhal, just outside the boundary of the STS, are extremely low at the present time with a median concentration of 4.4 Bq dm^{-3} (95% confidence interval: 4.1–4.7 Bq dm^{-3}). A subsequent measurement on electrolytically enriched well water from Sarzhal, sampled 2 years later in the course of a follow up visit, showed a tritium concentration of 3.9 ± 0.4 Bq dm^{-3} (Table IX), which is consistent with the median concentration observed in 2002, once radioactive decay is taken into account. Analysis of a few other Sarzhal well waters, also sampled in July 2004, yielded similar results (Table IX).

These levels are only slightly higher than the present background tritium content of surface and drinking waters globally. This is confirmed by the data summarised in Table V that show drinking water levels in various cities of Western Europe to have diminished to ~1 Bq dm^{-3} or less in recent years. As a benchmark, in the early 1950s, prior to the first period of heavy atmospheric weapons testing, tritium concentrations in US surface waters were estimated to be in or about 0.5 Bq dm^{-3} [5].

TABLE VIII. Tritium concentrations (±2σ) in domestic well waters sampled at the village of Sarzhal in July 2002 [Source: 24]

Source/code #	Co-ordinates	Depth (m)	Bq dm^{-3}
Well/ #01	49° 36' 08"; 78° 44' 33"	4	2.8 ± 0.9
Well/ #02	49° 36' 12"; 78° 44' 32"	6	3.8 ± 1.0
Well/ #03	49° 36' 24"; 78° 44' 24"	6	4.2 ± 1.0
Well/ #04	49° 36' 02"; 78° 44' 18"	4	4.3 ± 1.0
Well/ #05	49° 36' 02"; 78° 44' 13"	4	5.3 ± 1.0
Well/ #06	49° 36' 02"; 78° 44' 10"	7	4.4 ± 1.0
Well/ #07	49° 36' 08"; 78° 44' 31"	5	4.5 ± 1.0
Well/ #08	49° 36' 15"; 78° 44' 25"	5.5	4.5 ± 1.0
Well/ #9b	49° 35' 24"; 78° 43' 16"	3.5	4.3 ± 1.0
Well/ #30	49° 35' 54"; 78° 44' 07"	–	5.1 ± 1.0

Using, as baseline, the tritium concentration of 1.9 ± 0.3(2σ) Bq dm^{-3} recorded for Almaty Lake waters in July 2004, it can be inferred that about 40–50% of the tritium detected in Sarzhal well waters is of global fallout origin, with the remainder deriving from the conglomeration of tests conducted on the STS.

TABALE IX. Follow-up at Sarzhal 2 years on tritium concentrations in domestic well waters sampled at Sarzhal in July 2004 [The present study]

Source/code #	Co-ordinates	Depth (m)	Bq dm^{-3}
Well/#05	49° 36' 02"; 78° 44' 13"	4	<9
Well/#06	49° 36' 02"; 78° 44' 10"	7	3.9 ± 0.4
Well/#07	49° 36' 08"; 78° 44' 31"	5	<9
Well/#08	49° 36' 15"; 78° 44' 25"	5.5	<9

Clearly, to date, these wells have not been affected to any significant extent by contamination from underground testing conducted in the Degelen Mountains or at the neighbouring Balapan test field. However, it must be stressed that underground testing in the Degelen Mountains damaged the natural hydrological system in this zone, leading to unpredictable flushes of contaminated water from some of the many horizontal test tunnels in the area [19]. Following the banning of these tests the hydrological regime in the Degelen Mountains zone was gradually stabilised, with the sealing of tunnel portals and adits leading to a considerable decrease in the quantities of contaminated water released from these tunnels [19].

A similar picture to that for the Sarzhal wells emerges in the case of farm wells scattered throughout the eastern region of the SEMIRAD 1 study area. Although the number of wells in regular use is small, tritium levels are either close to 'background' or within an order of magnitude of 'background' in every case (Table X).

TABLE X. Tritium concentrations (±2σ) in well waters sampled on the STS in July 2002 [Source: 24]

Source	Code #	Co-ordinates	Depth (m)	Bq dm^{-3}
Degelen base:	Base well #10	49° 53' 03"; 77° 58' 17"		4.4 ± 1.0
Degelen base:	Tap-water #11	49° 53' 03"; 77° 58' 17"		3.5 ± 1.0
Tailan:	Farm well #12	49° 41' 09"; 78° 34' 05"	18	<1.7
Tailan:	Farm well #13	49° 40' 43"; 78° 37' 18"	2.5	52 ± 2
Shurek:	Farm well #14	49° 40' 48"; 78° 31' 59"	4	19.3 ± 1.3
Sholadir:	Farm well #26	49° 38' 30"; 78° 26' 20"		<1.7
Balapan:	Farm well #27	49° 57' 24"; 79° 04' 23"		21.2 ± 1.4

In some respects the data are surprising, as it has been speculated that water entering the study area from the Degelen Mountains is a major source of sub-surface water within the study area and of feed water to the Chagan River. Clearly, the Degelen streams are not a significant source of water present in either the farm wells or, for that matter, the Tel'kem craters, where the tritium concentrations are also very low. Moreover, the wells in Sarzhal village seem not to intercept waters flowing from Degelen. One possible explanation is that the waters flowing onto the study area from the Degelen Mountains mostly evaporate.

Alternatively, these groundwaters may recharge groundwater reservoirs that are not accessed by the relatively shallow wells examined here and that do not drain into the Chagan River (with one exception, the depth of the water table in these wells did not exceed 7 m). The above observations are supported by the fact that no evidence of enhanced concentrations of uranium, plutonium or americium attributable to underground testing was detected in these wells either [33].

4.5.2. The SEMIRAD 2 study area

Concentrations of tritium in well waters sampled on farms in that northern area of the STS lying to the south of the settlement of Maiysk (Fig. 3) in 2004 and 2005 were all extremely low and of no radiological significance whatsoever (Tables XI and XII, respectively). The same conclusion applies to pond and saltpan waters sampled in the SEMIRAD 2 study area during this period and also to water sampled alone the path of a stream meandering slowly from the southwest to the northeast across the study area.

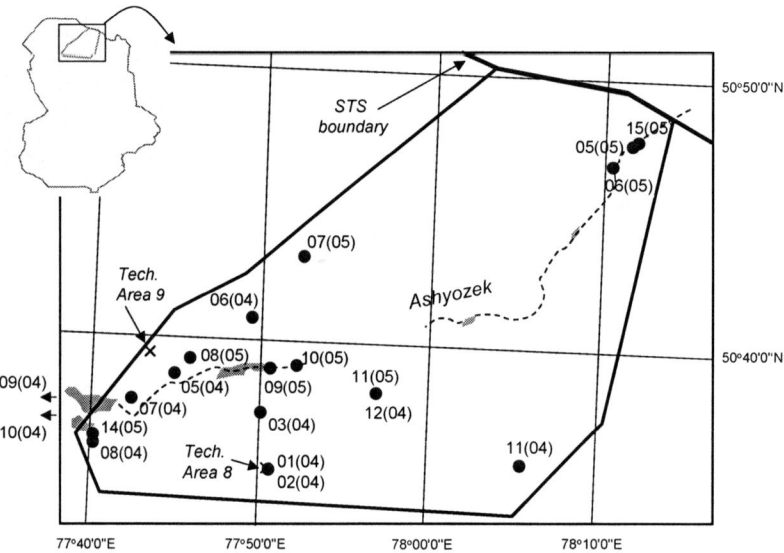

Figure 3. The SEMIRAD 2 study area showing the locations of streams, well waters and saltpans sampled in the July 2004 and July 2005 surveys

On the other hand, slightly elevated concentrations were recorded in water samples from two small man-made craters situated adjacent to one another in so-called *Technical Area 8* (Table XI). These craters are understood to have resulted from the testing of conventional explosive devices and this was confirmed by radiation survey monitoring of both craters in July 2004 (in the course of the SEMIRAD 2 project), which showed that neither crater was the result of a nuclear detonation.

4.5.3. *Radiological status*

The data confirm the relative absence of tritium in domestic wells in Sarzhal and in the SEMIRAD 2 study area, and very low levels in wells throughout the SEMIRAD 1 study area. In fact, many of the well waters examined show tritium concentrations that are indistinguishable from the mean global fallout content reported for water bodies of Russia, including Siberia in recent years [25, 26]. Moreover, the concentrations recorded in these wells are some three orders of magnitude below the guideline level for

TABLE XI. Tritium concentrations in well, lake and small crater waters south of Maiysk in the northern area of the STS (SEMIRAD 2, July 2004)

Source: code #	Co-ordinates	Tritium (Bq dm^{-3})
Farm well: #03(04)	50° 37' 20"; 77° 50' 04"	<6
Farm well: #05(04)	50° 38' 37"; 77° 44' 54"	<9
Farm well: #06(04)	50° 41' 02"; 77° 49' 20"	<9
Farm well: #08(04)	50° 36' 04"; 77° 40' 15"	<9
Farm well: #09(04)	50° 37' 37"; 77° 36' 38"	<9
Farm well: #10(04)	50° 36' 48"; 77° 22' 28"	<9
Farm well: #11(04)	50° 35' 45"; 78° 05' 23"	<9
Farm well: #12(04)	50° 38' 22"; 77° 57' 05"	<9
Crater (area 8): #01(04)	50° 35' 21"; 77° 50' 43"	30
Crater (area 8): #02(04)	50° 35' 21"; 77° 50' 43"	22
Small lake: #04(04)	Not recorded	<9
Small lake: #07(04)	50° 37' 45"; 77° 42' 34"	<9

TABLE XII. Tritium concentrations in well, salt-pan, stream and river waters south of Maiysk in the northern area of the STS (SEMIRAD 2, July 2005)

Source: code #	Co-ordinates	Tritium (Bq dm^{-3})
Farm well: #07(05)	50° 43' 17"; 77° 52' 28"	<6
Farm well: #11(05)	50° 38' 23"; 77° 57' 05"	<6
Farm well: #14(05)	50° 36' 12"; 77° 40' 09"	<6
Farm pond: #8(05)	50° 39' 01"; 77° 45' 41"	<6
Salt-pan: #09(05)	50° 38' 56"; 77° 50' 31"	7
Salt-pan: #10(05)	50° 39' 05"; 77° 52' 16"	<6
Salt-pan: #13(05)	50° 33' 28"; 77° 51' 55"	14
Ashyozek stream: #05(05)	50° 47' 38"; 78° 11' 38"	2.6
Ashyozek stream: #06(05)	50° 46' 32"; 78° 10' 42"	1.3
Ashyozek stream: #15(05)	50° 47' 39"; 78° 11' 36"	<6
Irtysh river: #01(05)	50° 27.866'; 79° 48.152'	<6
Irtysh river: #02(05)	50° 36.117'; 79° 25.904'	<6
Irtysh river: #04(05)	50° 36.416'; 79° 01.249'	2.5
Irtysh river: #12(05)	50° 45' 25"; 78° 33' 05"	<6
Near Chagan/ Irtysh confluence: #03(05)	50° 36.573'; 79° 16.336'	6.5

tritium in drinking water of 10 kBq dm^{-3} recommended by the World Health Organisation [34], and only marginally higher than the concentrations reported for drinking water in a number of other countries (Table V). The guideline level referred to here is based on a recommended reference level of committed effective dose of 0.1 mSv from 1 year's consumption of drinking water.

Clearly, the Sarzhal wells have not yet been affected by surface or sub-surface transport of radionuclides from test tunnels in the Degelen Mountains or vertical shafts (boreholes) at the Balapan test field adjacent to the SEMIRAD 1 study area. On the other hand, elevated levels of test-produced tritium were detected in stream waters entering the study area from the Degelen Massif and there is evidence that some of this tritium is being exported from the area via the Chagan River and other small tributaries.

Accordingly, it may only be a matter of time before the wells in question are affected unless, as postulated above, the complex hydrology of this geo-technically disturbed area, allied to high rates of evapotranspiration, preclude such an outcome. Regular on-going monitoring of the radiological status of these wells and streams is, nevertheless, to be strongly recommended. This advice is explored from the practical stand-point below.

5. Future Perspectives

5.1. THE BALAPAN TEST FIELD AND THE CHAGAN RIVER CATCHMENT

At the Balapan test field area, groundwater sampled recently from an unused borehole (Code #1419) showed a tritium concentration of 1.4 MBq dm^{-3} and a ^{90}Sr concentration of 2 kBq dm^{-3} [35]. Given that the nearest underground test was carried out 1 km away, it is clear that groundwater is currently mobilising radionuclides at the former test field.

It is also evident from the data discussed above that tritium is being exported outside the STS boundary, via the Chagan, from the test field area. The very high tritium concentration (281 kBq dm^{-3}) recorded in July 2004 6 km downstream of Balapan Lake is cause for concern, as are reports that relatively high tritium concentrations have occasionally been recorded in the Chagan River close to the Balapan test field in previous years. In fact the level of 281 kBq dm^{-3} is similar to the levels recorded in streams flowing from tunnel portals in the Degelen Massif. It should, however, be stressed that the presently available spatial and temporal data are few and, by virtue of same, inadequate, given the potential for transfer of tritium and, by extension, other radionuclides (e.g., ^{90}Sr, ^{137}Cs, ^{99}Tc) from the Balapan test field, where relatively high yield underground tests were conducted.

We, accordingly, advise that a renewed effort be made to gather a coherent set of data on tritium, ^{90}Sr, ^{137}Cs, ^{99}Tc and ^{239,240}Pu levels in the waters and associated sediments of this system. As said above, the IAEA has recommended that studies be conducted on the STS to investigate the status of local sources of drinking water, in order to determine whether these waters had been contaminated with radioactivity, particularly from tests carried out underground. This recommendation has been endorsed in an editorial on the subject of the STS [36], which appeared last year in the Journal of Radiological Protection. It is, therefore, gratifying to have had it reiterated at the outset of this workshop that one of the priority tasks of the NNC RK in the near future are tritium studies of groundwater dynamics at the STS. The case for detailed studies of the mobility of tritium in surface and ground waters in the Chagan River catchment is compelling. Ideally, for systematic hydrological studies, the issue of drilling new boreholes down to and below the water table, or cleaning out old boreholes around subterranean crushed zones/explosion cavities, should be addressed. Unfortunately, it is very costly to drill new boreholes. For example, Smith et al. [3], who drilled down to and below the water table (800 m) at the Nevada Test Site, estimated the cost at in excess of 10^6 USD per borehole (including well installation).

In the meantime, a less expensive interim strategy should be designed and implemented as soon as possible. It is suggested that the shape of the study be along the lines summarised in Table XIII below. Note the emphasis on (i) sampling along the length of the Chagan River with good spatial resolution; and (ii) sampling as many domestic/farm wells as possible within the catchment. To assist in defining and characterizing the

TABLE XIII. Proposed sampling scheme for the Chagan River catchment

Location	Sample	^{3}H	^{90}Sr	^{99}Tc	^{137}Cs	^{239,240}Pu
Chagan river & tributary:						
49° 52' 41"; 78° 44' 11"	River	√	√	√	√	√
49° 48' 01"; 79° 00' 07"	River	√				
49° 56' 14"; 79° 00' 30" (Balapan crater)	Atomic lake	√	√	√	√	√
49° 56' 31"; 79° 01' 23"	River	√	√		√	√
49° 59' 14"; 79° 02' 53"	River	√	√	√	√	√
50° 00' 37"; 79° 05' 28"	River	√	√		√	√
50° 12' 37"; 79° 20' 13"	Stream	√				
50° 23' 28"; 79° 17' 20"	River	√	√			
50° 32' 14"; 79° 16' 38"	River	√	√			
50° 36' 34"; 79° 16' 20"	River	√	√	√	√	√
Sources of drinking water:						
50° 21' 33"; 79° 12' 50"	Reservoir	√	√	√	√	
50° 01' 35"; 79° 08' 24"	Well	√	√			
49° 57' 32"; 79° 04' 55"	Well	√	√	√	√	√
49° 56' 42"; 79° 06' 10"	Well	√	√	√	√	√
50° 06' 12"; 79° 12' 52"	Well	√	√			
50° 08' 21"; 79° 11' 21"	Well	√	√			
50° 06' 20"; 79° 18' 47"	Wells	√	√			
50° 10' 50"; 79° 15' 41"	Well	√	√			
50° 10' 08"; 79° 17' 49"	Well	√	√			

latter, use could be made of a GIS-supported topographical model of surface water flow-lines for the Chagan catchment, available courtesy of IRSN (France). Additionally, high sensitivity, low background liquid scintillation spectrometers should be used for sample counting; this should enable tritium measurements as low as 1–2 Bq/l without the requirement of prior electrolytic enrichment of individual samples.

A field expedition to carry out some of the proposed sampling should be organised as soon as possible. For this exercise, two teams of four people in separate 4 × 4 vehicles should suffice, and should be able to complete the required sampling in the course of 2 or 3 days. However, it is desirable that similar sampling be conducted at different times of the year, so that seasonal variations can be identified and quantified. In relation to radiochemical analysis and measurement, the burden of work could be shared by INP/IRSE and Al-Farabi University, with support from IRSN-France (flow-line generation and deuterium analysis) and UCD-Ireland (electrolytic enrichment of low activity tritium samples).

5.2. GENERAL ISSUES

Other radionuclides of potential radiological significance in the longer term include ^{14}C, ^{36}Cl and ^{129}I, for which there are virtually no published data available for the STS. Both ^{14}C are ^{36}Cl are beta emitters and are readily amenable to low-level liquid scintillation counting, provided they have first been separated and purified using standard radiochemical techniques. At a minimum, both should be monitored in selected well waters and groundwaters.

Iodine-129 is also important as, being very long-lived (half-life = 15.7 My), it can be used to retrospectively reconstruct the pattern of short-lived 131I deposition on soils surrounding settlements in the vicinity of the STS, which arose from above ground weapons testing prior to the Limited Test Ban Treaty of 1963. For this purpose, it too should be measured in organic rich soils at these populous localities, preferably by accelerator mass spectrometry. The efficacy of this approach has been demonstrated by Straume et al. [37], who retrospectively reconstructed 131I deposition following the Chernobyl reactor accident of 1986.

6. Acknowledgements

This work was supported by the NATO Science for Peace programme (SEMIRAD Project, Contract SfP-976046(99) and SEMIRAD 2 Project, Contract SfP-980906(04)) and the EC Fifth Framework Programme (ADVANCE Project, Contract FIGE-CT-2000-00108). We also thank colleagues at the Institute of Nuclear Physics (INP), Almaty and the Institute of Radiation Safety and Ecology (IRSE), Kurchatov, for their advice and logistical support while on station in July 2002, 2004 and 2005.

References

1. Li-Xing, Z., Ming-shun, Z. and Gou-Rong, T. (1995) A field study of tritium migration in groundwater. *Sci. Tot. Environ.* **173/174**, 47–51.
2. Hoffmann, D.C., Daniels, W.R., Wolfsberg, K., Thompson, J.L., Rundburg, R.S., Fraser, S.L. and Daniels, K.S. (1983) A review of a field study of radionuclide migration from an underground nuclear explosion at the Nevada test site. *IAEA-CN* **43**, 469.
3. Smith, D.K. (1998) A recent drilling programme to investigate radionuclide migration at the Nevada Test Site. *J. Radioanal. Nucl. Chem.* **235**(1–2), 159–166.
4. Stead, F.W. (1963) Tritium distribution in ground water around large underground fusion explosions. *Science* **142**(3596), 1163–1165.
5. NCRP (1979) Tritium in the Environment, Published by the National Council on Radiation Protection and Measurements, Washington, DC. NCRP Report No. **62**, pp. 125.
6. Akhmetov, M.A., Artemev, O.I., Ptitskaya, L.D. and Sinyaev, V.A. (2000) Radiation monitoring of water flows and rehabilitation of Degelen mountain massif at Semipalatinsk test site, *Radioecology and Environment Protection*, NNC RK Bulletin, Kazakhstan, Issue 3, September 2000, pp. 23–28.
7. Dubasov, Yu. V. (2002) Radionuclides migration from nuclear testing tunnels in Degelen mountain of the former Semipalatinsk Test Site. *Proc. International Conference on Radioactivity in the Environment*, Monaco, 1–5 September 2002, pp. 290–295.
8. Shkolnik, V.S. (2002) The Semipalatinsk Test Site: Creation, Operation and Conversion, Sandia National Laboratories, *SAND* 2002-3612P, pp. 396.
9. Izrael, Yu.A. (2002) Radioactive fallout after Nuclear Explosions and Accidents, *Radioactivity in the Environment Series*, Amsterdam, Elsevier Science, **3**, pp. 281.
10. Matuschenko, A.M., Tsyrcov, G.A., Chernyshov, A.K., Dubashov, Yu.V., Krasilov, G.A., Logachov, V.A., Smagulov, S.G., Tsaturov, Y.S. and Zelentsov, S.A. (1998) Chronological list of nuclear tests at the Semipalatinsk Test Site and their radiation effects, in: Nuclear Test: long-term consequences in the Semipalatinsk/Altai Region. C.S. Shapiro, V.I. Kisilev and E.V. Zaitsev (eds.), NATO ASI Series, Environment. Springer, **36**, pp. 89–97.
11. Doriglasov, V., Dubasov, Yu., Dubkov, Y., Duvnik, V., Krasilov, G.A., Logachev, V., Maltsev, A., Matuschenko, A., Safronov, B., Smagulov, S., Stepanov, Y. and Tsaturov, Y. (1994) The Semipalatinsk and Northern Nuclear Test Sites of the USSR. General Characteristics of Releases and Depositions and Comprehensive Programme of Investigations of Radiological Impact of Nuclear Tests on Surrounding Territories. Abstract, Paper, Corrections and Addendum (V94-III) presented at NATO/SCOPE RADTEST Advanced Research Workshop, Vienna, Austria, 10–14 January, 1994.
12. IAEA (1998) Radiological Conditions at the Semipalatinsk Test Site, Kazakhstan: Preliminary Assessment and Recommendations for Further Studies. *Radiological Assessment Report Series*, International Atomic Energy Agency, Vienna, pp. 43.
13. Priest, N.D., Burkitbayev, M., Artemyev, O., Lukashenko, S. and Mitchell, P.I. (2003) Investigation of the Radiological Situation in the Sarzhal Region of the Semipalatinsk Nuclear Test Site. *NATO SEMIRAD Project Final Report*, Contract SfP-976046(99), February 2003, pp. 103.
14. Gordeev, S.K., Kvasnikova, E.V. and Ermakov, A.I. (2005) Radionuclide contamination of underground water and soils near the epicentral zone of cratering explosion at the Semipalatinsk test Site. *Radioprotection*, **40**(Suppl. 1), S399–S405.
15. Currie, L.A. (1968) Limits for qualitative detection and quantitative determination. *Anal. Chem.* **40**, 586–93.
16. Baeza, A., García, E. and Miró, C. (1999) A procedure for the determination of very low activity levels of tritium in water samples. *J. Radioanal. Nucl. Chem.* **241**(1), 93–100.
17. Taylor, C.B. (1977) Tritium enrichment of environmental waters by electrolysis: development of cathodes exhibiting high isotopic separation and precise measurement of tritium enrichment factors. *Proc. International Conference of Low-Radioactivity Measurements and Applications*, Slovenski Pedagogicke Nakladatelstvo, Bratislava, pp. 131–140.
18. Izrael, Yu., Stukin, E. and Ter-Saakov, A. et al. (1970) Radioactive contamination of natural environments from underground nuclear explosions and prediction methods. *Hydrometeoizdat*, Leningrad, pp. 67.(in Russian).
19. Konovalov, V.E., Pestov, E.Y., Artemjev, O.I. and Larin, V.N. (2000) Influence of the stabilization of hydrological regime on the ecology of the Degelen mountain (on the results of 1996/97 research).

Radioecology and Environment Protection, NNC RK Bulletin, Kazakhstan, Issue 3, September 2000, pp. 148–52.
20. Artemev, O.I., Akhmetov, M.A. and Ptitskaya, L.D. (2001) Radioactive contamination of former Semipalatinsk test site area. *Radioecology and Environment Protection*, NNC RK Bulletin, Kazakhstan, Issue 3, September 2001, pp. 12–19.
21. Kazachevskiy, I.V. et al. (2005) The results of radioecological investigations of the main surface water objects of the former Semipalatinsk nuclear test site'. *Proc. First International Nuclear Chemistry Congress*, Kusadasy, Turkey, 22–29 May 2005.
22. Subbotin, S. (2006) Radioactive contamination of groundwater at Degelen Site, STS. *Proc. NATO Advanced Research Workshop on Radiological Risks in Central Asia*, Almaty, 20–22 June 2006.
23. Kuzin, L.E. and Putilov, I.B. (2003) Short-term forecast of radioactivity carryover from tunnels with water manifestation at the Degelen Mountain Massif. *Proc. International Conference on Semipalatinsk Test Site – Radiation Heritage and Problems of Non-Proliferation*, Kurchatov, 7–9 October, 2003, Article No. 12, 6 pp. (in Russian)
24. Mitchell, P.I., León Vintró, L., Omarova, A., Burkitbayev, M., Jiménez Nápoles, H. and Priest, N.D. (2005) Tritium in well waters, streams and atomic lakes in the East Kazakhstan Oblast of the Semipalatinsk Nuclear Test Site. *J. Radiol. Prot.* **25**, 141–148.
25. Bolsunovska, A.Ya. and Bondareva, L.G. (2003) Tritium in surface waters of the Yenisei River Basin. *J. Environ. Radioact.* **66**, 285–294.
26. Makhonko, K., Kim, V., Kozlova, E., Volokitin, A., Mazurina, Z., Chumichev, V., Nikitin, A. and Katrich, I. (2001) Generalised data on radioactive contamination of natural environments. *Byulleten po Atomnoi Energii (Bulletin for Atomic Energy). TsNIIatominform* **10**, 26–32. (in Russian)
27. Pujol, L.I. and Sánchez-Cabeza, J.A. (2000) Natural and artificial radioactivity in surface waters of the Ebro river basin (Northeast Spain). *J. Environ. Radioact.* **51**, 181–210.
28. Hadzigehovic, M., Miljevic, N., Sipka, V. and Golobocanin, D. (1992) Environmental tritium of the Danube basin in Yugoslavia. *Environ. Poll.* **77**(1), 23–30.
29. EC (2001) Environmental radioactivity in the European Communities 1994. M. De Cort, S. Valdé, J. Van 't Klooster and G. Ponti (eds.), Office for Official Publications of the European Communities, EUR 18663 EN, 80 pp.
30. Villa, M. and Manjón, G. (2004) Low-level measurements of tritium in water. *Appl. Radiat. Isot.* **61**, 319–323.
31. Howlett, J.G. (2006) Personal communication.
32. Golikova, N.V., Artemyev, O.I., Larin, V.N. and Dontsova, G.A. (2003) Study of natural sorbents of Semipalatinsk region. *Radioecology and Environment Protection, NNC RK Bulletin*, Kazakhstan, Issue 3, September 2003, pp. 61–64.
33. León Vintró, L., Mitchell, P.I., Omarova, A., Burkitbayev, M., Jiménez Nápoles, H. and Priest, N.D. (in press) Americium, plutonium and uranium contamination and speciation in well waters, streams and atomic lakes in the Sarzhal Region of the Semipalatinsk nuclear test site, Kazakhstan. *J. Environ. Radioact.*
34. WHO (2003) World Health Organisation's Guidelines for Drinking Water Quality Third Edition, Chapter 9 (Draft), 17 February 2003.
35. Smith, D.K., Knapp, R.B., Rosenburg, N.D. and Thompson, A.F.B. (2003) International Cooperation to Address the Radioactivity Legacy in States of the Former Soviet Union, Preprint, UCRL-JC-154149, Lawerence Livermore National Laboratory and US Department of Energy (available electronically at http://www.doc.gov/bridge).
36. Grosche, B. (2005) Progress in assessing the public health impact from residues of nuclear bomb testing in Kazakhstan. *J. Radiol. Prot.* **25**, 123–124.
37. Straume, T., Marchetti, A.A., Anspaugh, L.R., Khrouch, V.T., Gavrilin, Y.I., Shinkarev, S.M., Drozdovitch, V.V., Ulanovsky, A.V., Korneev, S.V., Brekeshev, M.K., Leonov, E.S., Voigt, G., Panchenko, S.V. and Minenko, V.F. (1996) The feasibility of using 129I to reconstruct 131I deposition from the Chernobyl reactor accident. *Health Phys.* **71**(5), 733–740.

CHAPTER 6

SAFE MANAGEMENT OF RESIDUES FROM FORMER URANIUM MINING AND MILLING ACTIVITIES IN CENTRAL ASIAN IAEA REGIONAL TECHNICAL COOPERATION PROJECT

P.W. WAGGITT
Waste Safety Section, International Atomic Energy Agency, Wagramerstrasse 5, A-1400, Vienna, Austria

Abstract: Several of the Central Asian countries of the former Soviet Union were involved in the uranium mining and milling industry from about 1945 for varying periods until the break up of the Soviet Union in 1991. Even before the break up, several of these facilities had been abandoned and in only a few cases had any significant remediation been undertaken. Since 1991 the newly independent states of the region have been seeking assistance for the remediation of the multitude of tailings piles, waste rock stockpiles and abandoned, and often semi-dismantled, production facilities that may be found throughout the region. Many of these sites are close to settlements that were established as service towns for the mines. Most towns still have populations, although the mining industry has departed. In some instances there are cases of pollution and contamination and in many locations there is a significant level of public concern. The IAEA has been undertaking a number of Technical Cooperation (TC) projects throughout the region for some time to strengthen the institutions in the relevant states and assist them to establish monitoring and surveillance programmes as an integral part of the long term remediation process. The IAEA is liaising with other agencies and donors who are also working on these problems to optimise the remediation effort. The paper describes the objectives and operation of the specific TC regional programme, liaison efforts with other agencies, the achievements so far and the long term issues for remediation of these legacies of the "cold war" era.

Keywords: uranium mining and milling, monitoring, surveillance, IAEA regional programme

1. Introduction

The uranium mining activities of the former Soviet Union were widespread with many mining operations being located in the Central Asian Republics of Kazakhstan,

Kyrgyzstan, Tajikistan and Uzbekistan. Most of these operations commenced in the mid to late 1940s, but by the time of the break up of the former Soviet Union in 1991 few were still in operation and many of these mining and processing sites ceased activity at that time. In some cases there was little or no remediation of the site or waste disposal facilities and, with the downturn in economic activity that took place, some locations many sites were simply abandoned. The sites potentially contain a range of hazards to the environment and the population in the areas surrounding them. These may be physical (old buildings, open mine workings, pits and tunnels, derelict buildings and machines, etc.); chemical (acid drainage from reactive waste, old processing chemicals and residues); and/or radiological (tailings, unprocessed uranium-bearing ore, scale and sludge in old plants, contaminated scrap metal etc.). The Member States concerned do not have adequate access to the significant financial and other resources that are going to be required to evaluate the extent and nature of the problems as well as determining appropriate remedial actions and preparing work plans and actually undertaking the necessary works. The legacies that remain from former uranium mining operations have been catalogued for the most part, but many details still remain to be obtained before the final remediation objectives can be achieved. In particular the establishment of an appropriate institutional infrastructure in each member state has to be completed. At the present time there are varying levels of achievement in this area. Also as some Member States have only basic support infrastructure in the form of laboratories and field equipment their abilities to implement systems based on international standards are hampered.

All four of the countries mentioned are Member States of the International Atomic Energy Agency (IAEA) and have requested assistance from IAEA under the terms of the Agency's Technical Cooperation programme (TC). The IAEA assessed the situation and devised a Regional Project, which commenced in 2005, with the title "Safe Management of Residues from former Uranium Mining and Milling Activities in Central Asia". The rationale behind the establishment of a regional project was that all four Member States had similar issues relating to abandoned uranium mining and processing facilities within an area of similar climate and close together geographically. Thus to have such a project, which would encourage interaction and knowledge and experience sharing between the four Member States whilst also working to strengthen the local institutions and improve resources, offered IAEA an opportunity for improved service delivery throughout the region. In addition to the regional project two of the member States, Kyrgyzstan and Tajikistan, had previously each had a related national project accepted for implementation. The implementation of these two projects is being undertaken in such a way as to ensure that the activities were complimentary.

The project area is also a focus of assistance activity from other agencies who are working in the region. One significant activity has been to maintain liaison with these other agencies to ensure that there is a minimum of overlap between activities and optimisation of the combined efforts. In particular information relating to the areas of training and equipment supply has been shared between agencies. These activities are discussed later.

2. The Regional Project

The objectives of the Regional Project are (within each participating Member State):
- To develop a regulatory framework and decision making process to assess radiological impact and radiological residues at former uranium mining and processing sites
- To evaluate the remediation works underway
- To ensure international safety standards are being met
- To develop a plan of action to minimize the impact of radioactive residues on the population and assist sustainable development.

There are essentially four main mechanisms within the project programme that are being employed to achieve these objectives. These are workshops, training activities, equipment supply and scientific visits.

The major component is the series of four workshops which are spaced more or less evenly over the 2 years of the project's life so that each Member State will be host to one workshop. The workshops have been structured to introduce the concepts of broad assessment of the existing situations, planning activities, data collection, setting up monitoring and surveillance systems and their implementation, evaluation and justifycation of monitoring and surveillance, data assessment and reporting. Throughout the workshops there has been an emphasis on the need to employ international standards and much use has been made of the relevant IAEA documentation. In particular the IAEA Safety Guide on the management of residues from mining and processing of radioactive ores [1] and the IAEA Safety Report on the monitoring and surveillance of uranium mining and milling waste [2] have both been used as standard texts in all the workshops to date.

The initial workshop in Dushanbe (June 2005) provided an opportunity for the participants to describe the history of uranium mining and the associated legacies in each of their home countries and then to discuss the similarities of common problems. At this workshop the emphasis was on scene-setting in terms of the existing situation, information exchange on mutual areas of interest and common equipment requirements for implementation of the tasks necessary to work towards the project objectives. Most of the participants in the workshops have been senior and middle managers of the regulatory organizations involved with additional participation from those organizations with responsibility for the operation and remediation of the legacy sites. In this way the obligations and responsibilities of both regulators and operators can be addressed and explained in a common forum, which provides the opportunity for a full discussion between the parties involved.

During the workshops the participants have been given opportunities to practice what has been presented in the workshops by means of both practical exercises and assignments. In addition, group excursions to legacy uranium mine waste sites have provided real scenarios for discussion and evaluation "on the ground" where everyone is observing the same issue in the same location under the same conditions. This has provided the opportunity for debate without the risk of any party misunderstanding the site descriptions provided by others.

The second component in the programme is the provision of training enabling supervisory technical staff to raise their level of skills for tasks involved in the implementation of surveillance and monitoring tasks. This work has also provided the opportunity to introduce participants to the various items of equipment that are being supplied under the aegis of the project as well as reinforcing existing skills with existing equipment. The emphasis has been on establishing and maintaining networks of surveillance of monitoring around mine waste legacy sites. The main discussion point has been to ensure that participants understand how to make decisions on the methodology of such programmes; the selection of sampling locations and appropriate sampling methodology; the selection of analytical methods; data recording and analysis; the reporting and archiving of data; and the use of data in preparation of remediation plans and for provision of assurance to the public. In addition the need for quality control in all these activities has been emphasized.

The third component has been the supply of basic equipment to enable the participants and their colleagues to upgrade the level of performance and service that they can achieve in their respective home locations. The range of equipment has included material for both laboratory and field activities. Whilst some items supplied have been common to all groups there has also been an opportunity to provide items to address site or situation specific needs. Particular attention has had to be paid to ensuring that the equipment is not only suitable to the task to be undertaken but also the environmental conditions that may be encountered. These mining sites are generally associated with a wide range of climatic conditions which can be very harsh.

The fourth component has been the organization of scientific visits for some of the participants to observe what has been achieved elsewhere in the remediation of former uranium mining and processing facilities. The project has been able to establish a good working relationship with the Wismut company in Germany. This organization is responsible for the remediation of all the former uranium mining and processing facilities in the States of Thuringia and Saxony in eastern Germany. The Wismut company inherited a mining legacy which included included several open pits, about 1,400 km of underground mine workings, 311 million cubic metres of waste rock and 160 million cubic metres of radioactive tailings. All of these residues were located in or adjacent to populated areas. Thus the challenges faced by Wismut are comparable to the situations in many locations within the project Member States. The project is planning that two groups of personnel from regulatory organizations and two groups from operators will have the opportunity to visit the Wismut operation for a week. The first two groups, one comprising staff from regulatory bodies and one comprising staff from operating organizations with responsibility to remediate mining waste sites, have successfully completed their visits. The second round of visits are planned for September 2006, again one for regulatory staff and one for operating company staff. The programmes include not only visits to the sites which offer examples of remediation activity both ongoing and completed, but also briefing meetings with the local regulatory authority and opportunities to see all aspects of monitoring and surveillance activities actually being undertaken by Wismut staff. The visit programme also includes discussion sessions to explain the planning and review processes in maintaining monitoring networks and the importance of data analysis, storage, retrieval and reporting.

3. Liaison Activities

As previously mentioned the IAEA is not the only organization working in the region, undertaking assistance activities on the issue of safety in relation radioactive waste from uranium mining and processing, especially at legacy sites. A major programme has been the Environment and Security Initiative (ENVSEC) being implemented in the Ferghana valley by a group of agencies [3]. The members of the group are the Organisation for Security and Cooperation in Europe (OSCE), the United Nations Development Programme (UNDP), the United Nations Environment Programme (UNEP) and the Science for Peace programme of the North Atlantic Treaty Organisation (NATO). ENVSEC has several components but one is specifically dealing with the question of uranium mining waste safety and disposal. This programme contains elements of training and equipment supply which obviously have the potential to be very similar to the same activities in the IAEA project. In order to avoid nay unnecessary duplication of effort which would lead to a less than optimal use of the combined resources being applied to the resolution of these issues in the region an informal system of liaison has been implemented. This will be described later.

Another agency working in the region is the World Bank which has begun implementation of the Kyrgyz Republic Disaster Hazard Mitigation Project [4]. A significant part of this project will be the improved management and long term stabilisation of uranium mill tailings in the area of Mailuu-Suu in western Kyrgyzstan. Again the overall project plan contains components relating to technical training and the provision of surveillance and monitoring equipment. Again there is the potential for a duplication of effort and thus the potential for a less than optimal use of resources.

Another agency working on related issues has been the Lawrence Livermore National Laboratory from the United States of America, which has been providing funds and coordinating resources for low cost, smaller scale remediation of uranium mill tailings sites at Kadji Say, in another part of Kyrgyzstan.

In order to promote information exchange and to demonstrate to the Member States involved that efforts are being made to coordinate these various activities the IAEA has taken some initial steps in the area of liaison. At each workshop in the IAEA project has extended an invitation to each of these other agencies to attend and make a presentation to update all parties on the progress of their respective programmes. Each agency has been involved in at least one of the meetings and this has ensured that the participants can better understand how the various activities are all fitting together. Outside the workshops the agencies have been exchanging information on equipment supply and training activities to ensure that the various elements of individual agency programmes do not overlap and are coordinated. This has prevented unnecessary duplication of equipment supply and so enabled the range of items that are being provided to be coordinated to better effect. In the training sphere details of the various courses have been exchanged, again to ensure that activities are complimentary and not duplicated. It has been proposed that some training activities may be timed to be consecutive to further enhance the learning opportunities.

There is also the liaison issue of coordination with the two ongoing national projects in the region. In Kyrgyzstan the project is entitled "Establishment of a Radio-ecological Monitoring and Surveillance Network". This project is designed to be nationwide but the initial stages are concentrating in the areas of the former uranium mines at about five locations. The radioactive wastes from the former uranium mining activities are the source of the contaminants that are the prime target of the initial monitoring network. In addition to equipment supply and scientific visits this programme includes, training in the form of a fellowship for a staff member to study in another country, and support missions from international experts to assist with the setting up of the field activities programme. The project in Tajikistan is entitled "application of International Safety Standards on the Management of Uranium Mill Tailings". Again the linkage to the regional project is strong and some of the local counterpart staff has been involved in both programmes, which reinforces the need for careful liaison and integration between the IAEA work as well as with the work of other agencies. Again the focus of activity has been on supply of equipment and expert missions to provide training and assistance in the initial establishment of new monitoring programmes designed to international safety standards.

4. Project Progress

To date the project has been able to maintain a good level of activity. The performance indicators for the project are:

- Meetings programme completed on schedule
- Immediate action plans formulated by the end of 2006
- Adequate regulatory structure in place by the end of 2006.

To date the first objective is on schedule for achievement with three workshops completed successfully so far and the final event arranged for September 2006. The drawing up of immediate action plans has been a little slow to get underway and may not be completely finished by the end of 2006, however each Members State involved has made progress in this area. The final objective is unlikely to be wholly achieved as the level of success in creating suitable institutions and their associated institutional and physical infrastructure has been uneven across the four Member States. The feedback from the workshops has been good and the effort put into preparing assignments both during and between the workshops has been satisfactory. The main constraints seem to be a lack of resources in the areas of support infrastructure. The delivery of some of the equipment supplied under the project has been delayed through issues with suppliers and transport but this is being resolved. The development and introduction of monitoring and surveillance system based on International Standards has begun and will continue. The issues of data collection, quality control, analysis and reporting are being addressed. The scientific visits have shown what has been achieved elsewhere, albeit with access to many more resources than are currently available to any of the member States involved in the project. However, it has provided an insight into process and indicated what the time scales involved in realistic timetables for remediation programmes can be. Training has begun but already it is clear that further training in both field techniques and laboratory work would be beneficial. This is especially true in the case of the new equipment that is being delivered under the project, where recipients need to be assisted to adjust to the methods employed with the new items.

5. Future Activities

The initial project was designed to run for the 2 years of the IAEA TC cycle 2005–2006. The need to establish the on-ground situation with initial familiarisation visits and the delays in equipment procurement have meant that the programme was delayed in the early stages. It has not been possible to make up for all the time lost in the beginning, nor has the rate of progress been as rapid as planned. For this reason whilst the project will be successful overall it will be unlikely to complete all the objectives by the end of 2006. An extension of the programme has been agreed subject to the necessary funding being available which would extend project life into the 2007–2008 TC cycle. A new scope for this extended project is currently under preparation.

In addition to the project work there is also a proposal to create a more formal liaison process between the Member States and all the agencies working in the region on issues related to safety and radioactive waste management. This arose from a formal request to the IAEA from one of the Members States. A set of terms of reference is being drafted for consideration by all parties, a process that will hopefully be completed within the life of the present project. In addition to the proposed extension of the original regional project there are likely to be new national projects in the Member States currently involved. This could be with not only IAEA but also the other agencies involved in the region at present and possibly even new participants, both agencies and perhaps other Members States. Thus the requirement for an ongoing liaison process is likely to remain for some time.

The participating Members States are being encouraged to develop the monitoring and surveillance systems in relation to their uranium mining legacy wastes as a major step in the process of preparing long term remediation action plans. These data and the subsequent analyses and reporting, will form a significant input to waste characterization activities, design studies and the preparation of investment proposals for consideration by donor organisations.

6. Summary

The IAEA has an active TC programme in Central Asia which currently includes a regional project looking at the issues surrounding the safe management of waste from former uranium mining and milling operations. The safe remediation of legacy operations is important to the future security and safety of the environment and the population. Also, as the world uranium market seems set for a renaissance, the project presents an opportunity to introduce modern international safety standards that could be employed in any future uranium mining operations, either at old sites or in new locations.

The present project is using a combination of workshops, training activities, scientific visits and equipment procurement in the course of achieving the set objectives. Progress of the project has been good, although delays in some equipment supply and the variations in development of regulatory infrastructure and processes between the four Member States has resulted in progress being less rapid than anticipated. There has been a conscious effort to keep the project focused on real outcome sand to avoid overlapping with the work of other agencies working in the same region on related issues.

The project has been considered for extension and, subject to funding requirements being met, the work is going to continue. A revised scope for the extended project is currently in preparation.

References

1. International Atomic Energy Agency (2002) Management of Radioactive Waste from the Mining and Milling of Radioactive Ores. *Safety Standards Series* No.WS-G-1.2, IAEA, Vienna.
2. International Atomic Energy Agency (2002) Monitoring and Surveillance of Residues from the Mining and Milling of Uranium and Thorium. *Safety Reports Series* No.72, IAEA, Vienna.
3. http://www.envsec.org/
4. World Bank. Kyrgyz Republic Disaster Hazard Mitigation Project - http://wbln0018.worldbank.org/ECA/ECSSD.nsf/ExtECADocbyUnid/2F4DD0088206262A85256DF60080C4FB?Opendocument

CHAPTER 7

REHABILITATION OF URANIUM MINES IN NORTHERN TAJIKISTAN

M.M. YUNUSOV, Z.A. RAZIKOV,
N.I. BEZZUBOV, KH.I. TILLOBOEV
*SE "Vostokredmet", Kalinina Str., Sogd Oblast,
Khudjand City, Tajikistan Republic*

Abstracts: In the article "The conception on uranium mines rehabilitation" of the State Enterprise "Eastern combine for rare metals" an analysis is given of the engineering-geological and radiating conditions of the tailings of uranium production. Here are given the general principles of the approach for rehabilitation of the uranium mine territory. To illustrate the efficiency of the proposed approach, a practical case is shown, the rehabilitation of tailings from the hydrometallurgical plant Gafurov in Tajikistan.

Keywords: uranium tailing, rehabilitation strategy, Tajikistan

1. Introduction

Within the last years intensive work on evaluation of the inventory and estimation of the uranium manufactures conditions have led to the necessity of obtaining data about what is needed for their rehabilitation. Various actions are required, depending on the objects and the general conditions:

- Preservation of currently opened dumps
- Strengthening and preserving insufficient protection shelter
- Repair the broken preserving protection
- Recultivation of tailings, to transfer the territory to economic use
- Redislocation of tailings to a new, safer place
- Preserving the tailings in the current conditions with the organization of perspective control.

Each of these actions demands statement of control inspections with a set of methods, hardware and methodical maintenance. The purpose of the present work is the estimation of the tailings condition and development of main principles on rehabilitation of the territories where tailings are situated.

A number of dangerous highly radioactive objects have for the last 50 years been situated within the three uranium provinces in the north of Tajikistan (settlement Adrasman, Taboshar and right bank of Khujand) as a result of the uranium mining and plant processing activity [1–3].

The first experimental hydrometallurgical plant on enrichment of uranium ores started to operate in the Gafurov town area in 1945 year, on the basis of the Mailicyiskii and Uigur (Kirgizia) deposits. The hydrometallurgical plant in Taboshar was constructed simultaneously. The experimental plant in Gafurov was closed in the beginning of 1950s. The ore extraction and processing in Taboshar were stopped in the beginning of 1970s. Uranium ore extraction and processing were carried out in the Adrasman settlement in the 1940s–1950s and in 1970s–1980s the uranium ore extraction was carried out on the right bank of Khujand (mountain Mogoltau).

From the middle of 1950s the main (up to 95%) part of the raw material was delivered for processing at the enterprise from other regions of the former USSR and from distant foreign countries. Therefore about 90% of processed ores wastes processing was associated with imported raw materials.

Thus, numerous dumps of base ores, tailings and wastes from the mining and technological activity at the State enterprise "Vostokredmet", represent a potential radiating danger to population and the environment. The location for the objects is shown on the scheme (Fig. 1).

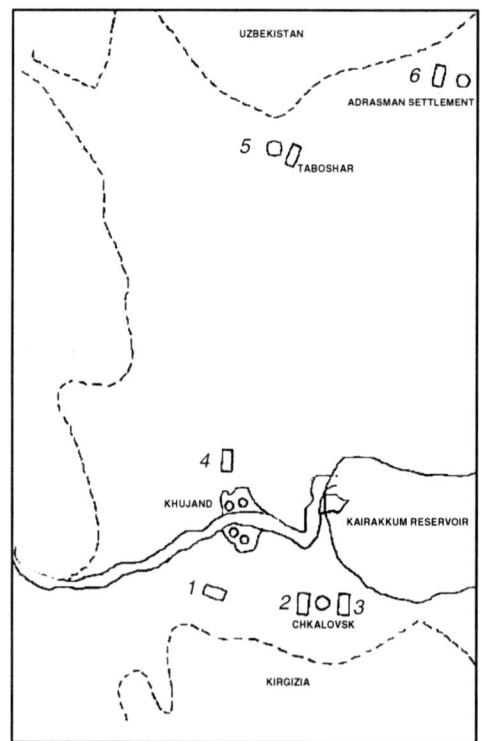

Tailing, right cost
[] Tailing, 1-Digmai tailing, 2-Tailing karta 1-9, 3-Gafurov tailing, 4-Tailing right bank, 5-Taboshar, 6-Adrasman

Figure 1. Scheme of tailing locations as a result of the work at SE "Vostokredmet"

Analysis of available material shows that the Taboshar non-buried tailings are the main sources of radioactive pollution. It should be noted, that the location of the radioactive objects is seismically dangerous (up to 8 points). Besides, during the last years there have been observed a substantial increase in the atmospheric precipitation and also their intensity. During the last years there have been two cases where tailings have been washed out by mud flows in Taboshar (tailings of the former shop 3) and two cases in Adrasman settlement (tailings 2) with partial washout of radioactive material and subsequent pollution of the adjoining district.

2. The Main Information About Tailings

The total waste quantity saved up during the period from 1945 to 1991 is about 54.8 million tons, situated in an area of 170 ha, with a total activity of 6.5 thousand Ci. The main data of tailings is given in Table I. The main work associated with recultivation and preservation of the tailings and dumps were carried out by the enterprise. The work was carried out in accordance with the norms of radiation-ecological safety used in 1970–1980s, and do not meet the requirements accepted in the territories of Russia and the CIS after 1991. The only object which applies to the modern burial international requirements is the Gafurov tailings where work was done under the project developed in 1992 by Ministry of Atomic Energy of Russia.

The condition of the radioactive objects is characterized as follows:

Chkalovsk city (natural gamma background is up to 20 µR/h):

- The tailings of Karta 1–9 is located in 1.5 km from the inhabited city sector. Exploited during 1949–1967, the area is 18 ha, the waste quantity is 3 million tons. Thickness of covering with neutral ground is 0.5–0.7 m. CED on a surface is 16–3,520 µR/h. Burial is required.
- The territory of Central Mine yard HMP was exploited till 1992. The area is 15 ha. CED is 60–80 µR/h. Tailings at HMP within a 90 ha area, the waste quantity is 19.5 million tons with total activity of stored wastes of 4,218 Ci, situated 1 km up on relief from large population centers and 9 km from Sir-Darya river. No expert controlling of the stability and the general radiation conditions of tailings, although the tailings are almost filled (83–85%) and situated in a seismic unstable area (up to 8 points).

Khujand, the right bank (natural gamma background is 45 µR/h):

- Edits dumps 1,2,2,3 former Mines 3, situated 4–5 km from the inhabited sector of the city which is down the slope in the foothills of the mountain Mogoltau. Exploited in 1976–1985, the total area is 59 ha, waste quantity is 0.35 million tons. Covering thickness with neutral ground is 0.5–0.7 m. CED on the surface is up to 30–60 µR/h. Drilling and blasting operations on the edits are required to prevent access for the public.

TABLE 1. Characteristics of the SE "Vostokredmet" tailings

Name of storing place and burial place	Tailings location and remoteness from the nearest settlement	Exploitation period	Sanitary-protective zone (m) area, hectare	Useful volume of tailing storehouse (m²) (numerator) and % its filling (denominator)	Isolation characteristics	Doze capacity in the zone of observation µR/h
1. Tailings	Digmai depression 1.5 km Gazien	From 1963	400 90.0	$19.4*10^6$ / 82.0	Opened	40–250
2. Tailings	Gafurov 0.5 km	1945 1950	– 4.0	$2.4*10^5$ / 100	Soil 2.5 m	20–60
3. Tailings	Karta 1–9 2 km Chalovsk	1949 1967	50.0 18.0	$2.6*10^6$ / 100	Soil 0.5 m	20–60
4. Tailings I–II	Taboshar 2 km	1945 1959	50.0 24.7	$9.88*10^5$ / 100	Soil 0.7–1 m	40–60
5. Tailings III	Taboshar 0.5 km	1947 1963	50.0 11.06	$1.06*10^6$ / 100	Soil 0.7–1 m	40–60
6. Tailings 4	Taboshar 1.0 km	1949 1965	50.0 18.76	$2.43*10^6$ / 100	Soil 0.7–1 m	40–60
7. Tailings workshop N3	Taboshar 3.0 km	1949 1965	50.0 2.86	$6.9*10^4$ / 100	Soil 0.7–1 m	40–60
8. Storage of base ores	Taboshar 4.0 km	1950 1965	– 3.35	$1.195*10^6$ / 100	Opened	40–100
9. Tailings N 2	1 km from Adrasman settlement	с 1991 г.	– 2.5	$2.4*10^5$ / 100	No	50–60
10. Mine -3 (4 bodies)	2 km from Khujand	1976 1985	– 5.9	$2.066*10^5$ / 100	Soil 0.5 m	60–80

* Note: the main polluting radionuclides in tailings' wastes: uranium, radium, polonium, thorium.

Taboshar (natural gamma background is 45 µR/h):

- Tailings from former Workshop # 3, located within 1 km from the inhabitant city sector. Exploited in 1949–1965, 2.9 ha area, waste quantity is 1.2 million tons. Covering thickness with neutral ground is 0.7–1.0 m. CED on a surface is up to 40–60 µR/h. Burial is required.
- Tailings of I–IV lines are located within 1.5–2 km from the inhabitant area. In case of emergency radioactive material washout is possible in all tailings in the Taboshar region downwards Utkonsy up to Syr-Darya River. Exploited I-II lines – 1945–1959, III line – 1947–1963, IV line – 1949–1965. Total area is 54.5 ha, waste quantity is 7.6 million tons. Covering thickness with neutral ground is 0.7–1 m. CED on a surface is up to 40–70 µR/h. Burial is required.
- Base ore plant dumps, mines 6 and edits 4 located in 2 km from inhabitant area are opened, and constant radioactive material wash out happens, also wind erosion occurs downwards the Sarimsahli with transport of materials to the Utronsi and further downwards to Sir-Darya River. Exploited in 1950–1965, area is 3.6 ha, wastes quantity is 2.2 million tons. CED on the surface is up to 300–500 µR/h. Project-survey is required for wastes burial.
- Low radwastes at the Sarimsahlisai river-bed, area is 10 ha, CED is up to 300 µR/h. Project-survey wok is required.

Adrasman settlement (natural gamma background is 45 µR/h):

- Tailings # 2, on the west outskirts of the settlement, in case of emergency the washout of radioactive materials from Karamazar can possibly be transported to Sir-Darya River. Founded in 1991 after performing of recultivating work in the region. Area is 2.5 ha, waste quantity is 0.4 million tons. Strewn with big fragmental material. CED is 40–60 µR/h.

The above-mentioned objects do not demand urgent work performance and work on their burial can be executed in future, but below-mentioned objects demand urgent works performance on their rehabilitation since they present a real radiating threat for the population and the nature.

The prime rehabilitation area are the Chkalovsk, Taboshar, Adrasman settlement due to their location in intermountain valleys in conditions of different relief (high mountains), the active display of exogenous physical-geological processes, and the high seismicity.

3. The Scope of Work on Rehabilitation of Polluted Territories

The locations of some tailings and dumps in the immediate proximity to rivers and streams, and also close to settlements together with the possibility of mudflow coming down, require not only radiometric research on objects but also detailed engineering – geological, hydraulic engineering and hydrogeological analyses of the natural-techno genetic situation of investigated areas. The result of the work is the substantiation of a rehabilitation programme associated with different objects and structures [3]:

1. Characteristics of the conditions for physical objects with evaluation of the radiation – ecological situation, calculation of the influence on the population with the forecast of the risks associated with man-caused and natural damages. The choice of recultivation will be based on the consideration of local and natural conditions.
2. The choice of direction of recultivation will be based on the provision for local nature and social conditions.
3. Binding of technologies to the concrete objects with expenses estimation.
4. Ranking of objects with respect to risk (danger degree) with development of a schedule of working projects on the concrete objects.
5. Estimation of expenses in the whole no all former uranium-mining objects in the republic with determining of prime works.

The purpose of the research is to make a decision to carry out the different item on the listed above actions. In the case if the full complex of research is necessary, the following must be characterized:

- Radiation condition of object surface
- Its influence on the environment (atmosphere, hydrosphere, biosphere, adjoining surface)
- Internal structure of the radiological object
- Data should be obtained to estimate the following factors of objects which may influence the environment
- Direct radioactive radiation
- Radon emission into the atmosphere
- Dispersion of radioactive and heavy elements as particles (dust formation, wash-out of the kept material from the surface by the time water-currents and etc.)
- Interrelation of tailing's material with underground waters.

At this stage the complex project will include gamma-shooting, emanation shooting, geochemical tailing material approbation, biogeochemical soil approbation and vegetation from the adjoining territory, hydrochemical approbation of underground waters on the observation holes. The radiating inspection will be carried out after the choice of a specific action.

Plan-periodic radiating control is being carried our during the action performance which has the purpose to estimate the efficiency of current work, correct their direction, make dosimetry of the personnel and techniques. In this case the structure of methods' complex is formed by the principle of necessary sufficiency.

After the performance of the planned actions to decrease or prevent the influence from an object on the environment, the program of long-term control will be made. The long-term control should be considered in two aspects – the control of an object and the radiation–ecological monitoring of the surrounding territory. In the first case, the control purpose is the prevention of corrosion, erosive and other damages of the protective barriers of the rehabilitated objects, development and performance of preventive actions to avoid the appearance of radioactive sources in adjoining territory and to make the radiation-ecological forecast.

4. Practical Realization of the Concepts

The same approach to the organization of the rehabilitation work has been used by us in the case of the tailings from the former experimental hydrometallurgical works in Gafurov; i.e. the category "buried" and tailings of the operating Digmai. For the Gafurov tailings all four stages of the radiating monitoring programme were used. The decision on its burial and the burial location place were based on the preliminary survey results.

At the second stage, sites in the adjoining territory being polluted with radioactive materials were identified, the material for shelter was chosen and its capacity was established, guaranteeing a lowering in the radiating characteristics to reach the background level.

At the third stage, environmental radiation control was carried out at the polluted sites after the re-deployment of the radioactive material; radiation control was carried out at the tailings during the burial. After fulfilling the rehabilitation action, the radiation conditions on the tailings is characterized by following data: gamma dose rate is within 12–15 μR/h, radon escalation does not exceed 0.04 Bq m^2/sec, which corresponds to the background measurements.

At the fourth stage (the long-term control) the estimation of the conditions for an engineering – technical tailing is carried out annually from 1992, the gamma-background and radon escalation are measured. The Digmai tailing is functioning. Filled to 80% and consequently its temporary closing-down are inexpedient. Therefore, we organized a long-term control over the tailing's condition and its influence on the environment of the adjoining territory. The gamma dose on the tailings surface iswithin the range 600–800 μR/h. Radon emission is estimated in 120 Ci/year. The radioactive contaminated areas are identified from the results of area gamma-shooting, within the limits of a protective sanitary zone. The map of flora is based on the results from botanical and biochemical research in 1991, and the control programmes for the subsequent measurements are chosen (Fig. 1). Based on electro-profiling by means of a vertical electric sounding, the area of polluted underground waters is determined and the directions of migration of the pollutants from the tailings are established. This work was repeated in 1994 and 2005. A change in the distribution of the vegetative cover has been observed, which can be related to the microclimatic conditions in the tailing area in connection with the disappearance of the pond.

The radioactive contaminated surface area in the northeast direction has increased, which was caused by water transport of tailing materials and dust over time. At present, the surface of the tailings has been sowed with cane to prevent dusting. For the forecast and the analysis of influence of the tailings, the package of applied programs "Ecology of underground waters" was developed and applied. The similar package is under development for the atmosphere. Its introduction will allow the analysis of the distribution of radon and dust from the tailings surface. We consider that the experience of the executed work will be useful for other tailings, particularly in Taboshar, Tajikistan.

5. Conclusions

On the base of the radiation conditions of the tailings, the complex of actions has been developed to perform and fulfill the rehabilitation of polluted areas such as tailings.

The research complex for each stage of work has been developed, which provides the choice of rehabilitation actions and controlling the obtained results.

To illustrate the efficiency of the proposed approach, the rehabilitation program for the Gafurovs' hydrometallurgical plant tailings is presented. The experience acquired from the rehabilitation work at this site will be useful for the rehabilitation of other tailings.

References

1. Govard A.D. and Remson I. (1978) Geology and protection of environment/translation from English under J.K. Burkova's edition. L. Nedra, 1982, 583 pages. First edition, USA.
2. Koshik J.I. (2005) Rehabilitation of territories, exposed to uranium activity manufactures. Report at the seventh session of the commission of the countries of participants of the CIS on use of an atomic energy in the peace purposes. Almaty, 35pages.
3. Soroka Yu.N., Kretinin A.V., and Molchanov A.I. (1993) Sanitary rules of liquidation, preservation and re-profiling of the enterprises on extraction and processing of radioactive ores (LKP-91) M, 1991, Atomic Energy, Springer, **75** (2), 654–659.

CHAPTER 8

A RATIONAL APPROACH TO BRIDGING THE NUCLEAR TECHNOLOGY USAGE AND NUCLEAR EDUCATION GAP

S.N. BAKHTIAR
Institute of Dynamic Change
2135 Ascot Dr. #28
Moraga, CA 94556

Abstract: Today's total world energy demand is nearly 200 million B/D of oil equivalent, up five-fold from 1950. Many forecast an imminent decline in oil and gas production, but population growth and economic development push demand upward. Dramatic changes must occur in energy supply and demand beyond 2050. Major problems exist in effectively capturizing, converting, storing, transporting and utilizing different forms of energy while meeting societies' diverse and changing economic, environmental, political, cultural, geographic and aesthetic needs. Development of technology, though difficult, is necessary and almost certainly achievable. The present paper presents a short review on past changes in technology and energy, future energy demands, future energy resources including nuclear energy and focuses on challenges in the educational arena.

Keywords: energy sources, energy use, energy demand, educational challenges

1. Introduction

The year 2050 is a 44-year leap into the future, so it is instructive to review past changes in technology, so it is instructive to review past changes in technology and energy use. In 1850, wood supplied 70% of world's 'commercial" energy. Steam engine use, fired by coal and wood, grow with the industrial revolution. Locomotives and horses transported most bulk goods and people on land. Agriculture, and most local transportation, was all muscle powered [1–3]. Wind was used almost exclusively at sea. Light came from fire, candles, whale oil, and "city gas" in urban areas. The military moved troops, arms, and supplied by horse, rail, or foot on land, and by wind at sea [3].

Globalization process expedited drastically since early twentieth century started with invention of electricity, automobile, airplane, telephone, television, computer, etc. Until this era, no one could imagine the global impacts caused by the revolution of transportation, communication, industrialization in our modern times. The amount of available information is exponentially growing. Information availability and control became unfamiliar problems to dictators and oppressed regimes, although the direct impacts of modern revolutions caused in many advancements in technologies, science, and medicine.

Today's total world energy demand is nearly 200 million B/D of oil equivalent, up fivefold from 1950. Many forecast an imminent decline in oil and gas production, but population growth and economic development push demand upward. Dramatic changes must occur in energy supply and demand beyond 2050 [4–5].

Yet vast resources of energy exist. About 1.4×10^{19} BTU of solar radiation hits Earth daily, 13,000 times current total energy use. Another 5 to 8×10^{14} BTU, roughly equal to current use, flows to Earth's surface from its interior. Every pound of material equals nearly 4×10^{13} BTU, so each barrel of oil contains more than two billion times more energy than is available by combustion [6].

As more and more developing countries industrialize, they will naturally want more energy to quench the growth thirst. This will see more involvement in international affairs, and indeed China and India are increasingly active in many regions around the world. When we think of the world, we tend to think of it as big, perhaps because we are unable to fully comprehend it in all of its complexities and diversity and exotic geography. We can read about some tragedy or noteworthy event with little more than a shrug because it happened "over there." But in reality, the world is getting smaller as people become more and more interconnected. Call it globalization, call it a communications revolution, and call it inevitable. The fact is we can no longer realistically expect to localize the issues and challenges we'll face in this century [7–8]. Geopolitical issues, new and old will therefore arise. Legitimate stability and supply issues are also of concern. For example, places like Nigeria, Iraq, Iran, etc. all produce oil, but present problems of varying degree for oil consuming nations, as concerns range from stable supply to stable government, and for others, such as Venezuela, to use oil and its related profits to develop their own country even more.

The modern western nations form a small percentage of the world population, but consume far more resources. Problems such as energy depletion are largely caused by these nations. However, as China and India also grow rapidly there is a fear that these countries' demands for energy and resources will vary very quickly so that the world's natural resources are stripped away even more quickly given their larger population sizes [4–6].

As mentioned earlier, today's total world energy demand is nearly 200 million B/D of oil equivalent, up five fold from 1950 – more than 80% supplied by fossil fuels, and nearly 60% by oil and gas. Many forecast an imminent decline in oil and gas production, but population growth and economic development such demand upward. Dramatic changes must occur in energy supply demand beyond 2050. Major problems exist in effectively capturizing, converting, storing, transporting and utilizing these forms of energy while meeting societies' diverse and changing economic, environmental, political, cultural, geographic and aesthetic needs. Development of technology, though difficult, is necessary and almost certainly achievable [6, 9–10].

In the IEO2006 reference case, the world's total net electricity consumption doubles, growing at an average rate of 2.7% per year, from 14,781 billion kilowatt hours in 2003 to 21,699 billion kilowatt hours in 2015 and 30,116 billion kilowatt hours in 2030. Non-OECD countries account for 71% of the projected growth and OECD countries 29% [6].

Projected growth in net electricity consumption is most rapid among the non-OECD economies of the world, with annual average growth of 3.9% from 2003 to 2030, compared with 1.5% for OECD economies. China and the United States lead the growth in annual net electricity consumption with increases of 4,300 and 1,963 billion kilowatt hours, respectively, over the projection period [9–10].

In the United States, electricity demand increases from 3,669 billion kilowatt hours in 2003 to 5,619 billion kilowatt hours in 2030. Demand growth in the commercial sector is particularly strong, averaging 2.2% per year. Additions to commercial floorspace, the continuing penetration of new telecommunications technologies, and increased use of office equipment offset efficiency gains for electric equipment in the sector. Moderate increases are projected for electricity consumption in the industrial and residential sectors, averaging 0.8% per year and 1.5% per year, respectively. A similar pattern is projected for Canada, where net electricity consumption grows from 521 billion kilowatt hours in 2003 to 660 billion kilowatt hours in 2015 and 776 billion kilowatt hours in 2030.

Electricity consumption in the non-OECD economies grows at an average annual rate of 3.9% from 2003 to 2030. Non-OECD Asia has the highest growth rate at 4.7% per year, followed by Central and South America at 3.7%, the Middle East at 3.0%, Africa at 2.9%, and non-OECD Europe and Eurasia at 2.8%. The average annual growth rates translate to a near tripling of net electricity consumption in the non-OECD nations over the projection period. In 2003, non-OECD economies consumed 40% of the world's electricity; in 2030 their share is projected to be 56%. Growth in net electricity consumption for the non-OECD economies is driven in large part by assumptions about GDP and population growth.

To meet the world's electricity demand over the 2003 to 2030 projection period, an extensive expansion of installed generating capacity will be required. In the reference case, worldwide installed electricity generating capacity grows from 3,710 GW in 2003 to 6,349 GW in 2030, at an average rate of 2.0% per year [10].

2. Natural Gas and Coal

At the world level, natural gas consumption increases from 19% of total fuel use for electricity generation in 2003 to 22% in 2030. Non-OECD economies, on the whole, relied on natural gas for 24% of fuel inputs in 2003 and OECD economies for 15%. No change is expected for the non-OECD economies, but in the OECD the natural gas share rises to 20% in 2030.

In the OECD economies, natural gas and coal each accounted for 28% of installed electricity generating capacity in 2003. Over the projection period, natural gas capacity gains share (rising to 33%) at the expense of nuclear, renewables, and oil-fired capacity, while coal's share remains steady. Nearly one-half of the total increment in OECD natural-gas-fired generating capacity is attributed to the countries of Europe, where the natural gas share of electric power generation more than doubles, from 15% in 2003 to 39% in 2030. With planned phaseouts of nuclear generators in Belgium, Germany, and Sweden and disincentives for construction of new coal-fired capacity because of environmental re-strictions, natural gas gains the largest share of the OECD Europe electricity market.

Coal retains the largest market share of the world's electricity generation (roughly 40%) in the IEO2006 reference case, despite losing some of its share to natural gas. Installed coal-fired capacity, as a share of total world capacity, remains at about 30%. Worldwide, coal-fired capacity grows by 2.2% per year, from 1,119 GW in 2003 to 1,997 GW in 2030 slightly faster than the 2.0% average annual increase for all electricity generation capacity. In 2003, non-OECD economies on the whole relied on coal for roughly 43% of generation, slightly more than the OECD economies.

Regional differences in coal use for electricity generation arise primarily from differences in coal resources. Regions with large coal resources are more likely to use coal for electricity generation, because coal has a lower energy density (energy per weight) and fewer alternative uses than oil or natural gas. These factors help keep coal prices, on an energy basis, lower than oil and natural gas prices. Coal reserves in the United States, China, India, and Australia are among the largest in the world, and those countries rely on coal to generate 50% to 80% of their electricity.

3. Oil-Fired Generating Capacity

Although relatively little change in oil-fired generating capacity is expected, oil's share of world installed capacity declines over the projection period, from 10% in 2003 to 7% in 2030. Oil has more value in the transportation sector and in limited applications for distributed diesel-fired generators than in central power plant applications. Only the Middle East and China are expected to see sizable increases in oil-fired electric power capacity over the projection period, adding 24 and 22 GW, respectively.

In recent years, China has shown fairly strong growth in oil-fired electricity generation, because peak electricity demand continues to outpace on-grid electricity generation, and Chinese industry has had to rely on diesel generators to cope with annual summer power shortages. That situation is expected to continue in the short term, but as planned capacity fueled by natural gas, coal, nuclear, and hydropower comes on line and the country's national electricity grid matures, the use of oil to generate electricity is expected to moderate.

4. Nuclear Energy

The world's nuclear-powered generating capacity increases in the IEO2006 reference case from 361 GW in 2003 to 438 GW in 2030, in contrast to projections of declines in nuclear power capacity in past IEOs. The reference case is based on existing laws and assumes that, for the OECD economies in the long term, retirements of existing nuclear power plants as they reach the end of their operating lives will nearly equal construction of new nuclear power capacity, resulting in a slight decline of installed nuclear capacity toward the end of the projection after peaking in 2020. Few new builds are expected in the OECD economies outside of Finland, France, Japan, South Korea, and the United States. In the United States, nuclear capacity is expected to increase by 3 GW as a result of uprates at existing plants and by 6 GW as a result of new constructions.

In contrast, rapid growth in nuclear power capacity is projected for the non-OECD economies. The non-OECD economies are expected to add 33 GW of nuclear capacity between 2003 and 2015 and another 42 GW between 2015 and 2030. The largest additions are expected in China, India, and Russia.

Prospects for nuclear power have improved in recent years, with higher capacity utilization rates reported for many existing nuclear facilities and the expectation that most existing plants in the OECD nations and in non-OECD Europe and Eurasia will be granted extensions to their operating lives. Higher fossil fuel prices, concerns about energy supply security, and the possibility for new, lower cost nuclear reactor designs also may improve prospects for new nuclear power capacity. Nevertheless, nuclear power trends can be difficult to anticipate for a variety of political and social reasons, and considerable uncertainty is associated with nuclear power projections.

Nuclear power is an important source of electricity in many countries of the world. In 2005, 16 countries depended on nuclear power for at least 25% of their electricity generation. As of December 2005, there were 443 nuclear power reactors in operation around the world, and another 24 were under construction. Despite a declining share of global electricity production, nuclear power is projected to remain an important source of electric power through 2030. In the IEO2006 reference case, electricity generation by nuclear power plants around the world increases from 2,523 billion kilowatt hours in 2003 to 2,940 billion kilowatt hours in 2015 and 3,299 billion kilowatt hours in 2030.

In the OECD, grid-connected installed renewable capacity is projected to increase by 0.8% per year over the 2003 to 2030 period. Hydroelectric capacity in OECD economies is not expected to grow substantially, and only Canada is expected to complete any sizable hydroelectric projects over the projection. Non hydropower renewable are instead expected to lead the growth in renewable generating capacity, especially wind in OECD Europe and the United States, where wind-powered generating capacity increased by 18% and 27%, respectively, in 2005 alone.

The IEO 2006 projections for hydroelectricity and other renewable energy resources include only on-grid renewable. Non-marketed (noncommercial) biofuels from plant and animal sources are an important source of energy, particularly in non-OECD economies, and the International Energy Agency has estimated that some 2.4 billion people in developing countries depend on traditional biomass for heating and cooking. Because comprehensive data on the use of non-marketed fuels and dispersed renewable (renewable energy consumed on the site of its production, such as solar panels used to heat water) are not available, they are not included in the projections; however, both non-marketed fuels and dispersed renewable are considered in formulating end-use energy demands.

5. Future Challenges

The present mix of fuels and energy technologies is not sustainable in the long term, even if population were dramatically reduced. In the long term, fossil fuels will run out. Even now our present ways of using energy have led to serious levels of pollution build up in almost every big city in America. Also attribute to toxic contamination of ground water and into natural bodies of water. Regardless of long-term sustainability, the growth of population and the composition of this growth in the next 2–3 decades are possibly the most serious energy crises. One of these challenges is the need to develop alternate energy sources. It is no longer a question of if we should; it's a question of when, and the answer is now. Subjecting future generations to the tenuous viability of the status quo is simply not an option. Many nations have already made the commitment to alternative energy – primarily nuclear – and are leading the way in this pursuit. Many others are now beginning to understand that this is the future, that it is imperative that we, as a global community, break free from our reliance on fossil fuel-based energy.

As energy plays a central role in the world development, it represents as well as major challenge for sustainable development. Even if technology developments will reduce specific consumption, the world energy demand is likely to increase in line with its population. Energy and material efficiency and the integration of the renewable resources will therefore have to play a major role for sustainable development.

The challenge concerns not only the technologies at the conversion and useful energy level, but the energy management and infrastructures. Of course, to get to that future, we must face some present-day realities – namely, the complex task of establishing cohesive, enforceable policies for nuclear energy management and assuring that these policies do not ultimately lead to the proliferation of nuclear weapons. In summary before discussing some of the challenges in the educational arena, it is very vital to emphasis on the following points:

1. A retrospective look at 150 years of human activities and technology and their combined impacts on types of energy and its use suggest that technology will have a significant and largely unpredictable impact on future energy needs and how they are met.
2. World population is expected to grow by 50% over the next 50 years, implying a 50% increase in energy needs in 2050 if consumption per capita remains constant.
3. Global economic development, considered a favorable objective, could add significantly to energy demand.
4. Geopolitics, environmental issues, and economics will impact the future world supply and the cost of fossil fuels over the long-term more than physical shortages.
5. Enormous supplies of renewable energy are available – solar, including wind and biomass, and geothermal – but their dilute and intermittent nature is constraining.
6. Nuclear power, particularly reprocessing of spent nuclear fuel and fast breeder reactors, has the potential for meeting most future electrical and other energy needs. When cost-effective fusion reactors become available, nuclear power will be essentially infinite.
7. Increasing efficiencies and the falling costs of photoelectric cells promise more use of direct solar power. In some areas, wind and hydropower are now cost competitive with new coal-fired electric generating plants.

8. Transportation fuel can be provided by electrically converting natural gas, coal, or biomass to liquids for either IC engines or fuel cells, and for producing hydrogen for both.
9. Strictly speaking there cannot be a "shortage of energy" in a free-world market. Price forces a balance between supply and demand. Higher prices, or the prospect thereof, also drive innovation and new technology, creating newer, cheaper, and better sources of energy and more efficient use of old ones.

6. Educational Needs

Another reality is the current shortage – in the U.S. as well as around the world – of highly-trained nuclear scientists and engineers. This, for obvious reasons, is problematic, as the demand for such expertise will only increase as we go forward in our efforts.

There is a quiet crisis building in the United States – a crisis that could jeopardize the nation's pre-eminence and well-being. The crisis has been mounting gradually, but inexorably, over several decades. If permitted to continue unmitigated, it could reverse the global leadership Americans currently enjoy.

The crisis stems from the gap between the nation's growing need for scientists, engineers, and other technically skilled workers, and its production of them. As the generation educated in the 1950s and 1960s prepares to retire, our colleges and universities are not graduating enough scientific and technical talent to step into research laboratories, software and other design centers, refineries, defense installations, science policy offices, manufacturing shop floors and high-tech startups [7–8].

This "gap" represents a shortfall in our national scientific and technical capabilities.

The need to make the nation safer from emerging terrorist threats that endanger the nation's people, infrastructure, economy, health, and environment, makes this gap all the more Critical and the need for action all the more urgent. We ignore this gap at our peril. Closing it will require a national commitment to develop more of the talent of all our citizens, especially the under-represented majority – the women, minorities, and persons with disabilities who comprise a disproportionately small part of the nation's science, engineering, and technology workforce.

The American public has not focused on the quiet crisis because we have grown accustomed to the fruits of technology. The technological advances of the past 100 years created a cornucopia of riches that have dramatically altered the quality and nature of daily life. Few Americans can remember life before electricity and electronics; ground, air, and space transport; radio and television broadcast; telephonics and satellite communications; medical technologies and imaging for diagnostics, treatment, prevention, and health assurance; laser and fiber optic, petrochemical, and nuclear technologies [6].

The U.S.-led surge in information technology that began in the early 1990s fostered a shared sense that prosperity could be taken for granted. Then-new technologies such as the World Wide Web, e-mail, and reasonably priced microprocessors boosted

American productivity and spread rapidly through most segments of the economy. Life for many Americans was comfortable, safe, healthy, convenient, relatively wealthy, and thoroughly endowed with choice and consistency. The golden continuity of prosperity – together with the break-up of the Soviet Union and the triumph of market-based economics – signaled a new millennium in which the foundation of U.S. strength could be assumed. The assumption of continued progress – even American invincibility – was shattered on September 11, 2001 [11]. The hard questions that have been asked since then have centered on the immediate capacity of the nation to fight terrorism. But the current natural focus on intelligence capabilities and defense preparedness should not overshadow the most fundamental of questions. Is the United States developing the human capital to remain the world's most productive economy while at the same time meeting a formidable new national security threat?

The Council on Competitiveness, which for 15 years has studied the capacity of the nation to support high-wage jobs and win in global markets, has shown how much scientific and technical talent contribute to national economic performance in a world where advanced knowledge is widespread and low-cost labor is readily available, U.S.

Advantages in the marketplace and in science and technology have begun to erode. A comprehensive and coordinated federal effort is urgently needed to bluster U.S. competitiveness and pre-eminence in these areas. This congressionally requested report by a pre-eminent committee makes four recommendations along with twenty implementation actions that federal policy makers should take to create high-quality jobs and focus new science and technology efforts on meeting the nation's needs, especially in the area of clean, affordable energy [12]:

1. Increase America's talent pool by vastly improving K-12 mathematics and science education.
2. Sustain and strength the nation's commitment to long term basic research.
3. Develop, recruit, and retain top students, scientists, and engineers from both the U.S. and abroad.
4. Ensure that the United States is the premier place in the world for innovation.

Some actions will involve changing existing laws, while others require financial support that would come from relocating existing budgets or increasing them.

References

1. U.S. Department of Commerce (2002), U.S. Summary: 2000, Census 2000 Profile.
2. Forest & Sullivan 2002 Revenue Forecasts.
3. Ricketts et al. (2005), Pinpointing and preventing imminent extinctions, Proceedings of the National Academy of Sciences of the U.S. 10.1073/pnas.0509060102, p.s
4. Energy Information Administration (EIA) (2002), International Energy Outlook 2002, DOE/EIA-0484, March 2002, p. 276.
5. Energy Information Administration (EIA) (2005), International Energy Outlook, 2005, DOE/EIA-0484, July 2005, p. 194.
6. Porter, M.E. et al. (2000), The Global Competitiveness Report 2000, World Economic Forum, Geneva, Switzerland, Oxford University Press, 2002, ISBN-0-19-513820-1, p. 332.

7. Nuclear Science Advisory Committee (2004), Education in Nuclear Science, A Status Report and Recommendation for the Beginning of the 21st Century, DOE/NSC Nuclear Science Advisory Committee – Subcommittee on Education, November 2004, p. 175.
8. IAEA (2002), Assessment of the Teaching and Application in Radiochemistry, Report-2002, IAEA.
9. Horton, R. (2003), World Energy beyond 2050, JPT Online, University of Wisconsin-Milwukee, December 2003.
10. Energy Information Administration (EIA) (2006), International Energy Outlook, 2006, DOE/EIA-0484, June 2006, p. 202.
11. Huesmann, M. H. (2006). Can Advances in Science and Technology prevent Global warming? Mitigation and Adaptions Strategies for Global Change, 11: 539–577.
12. Werbos, P. J. (1993), Energy and Population: Transitional Issues and Eventual Limits, Negative Population Growth – The NPG Forum.

CHAPTER 9

STRONTIUM-90 CONTAMINATION WITHIN THE SEMIPALATINSK NUCLEAR TEST SITE: RESULTS OF SEMIRAD1 AND SEMIRAD2 PROJECTS – CONTAMINATION LEVELS AND PROJECTED DOSES TO LOCAL POPULATIONS

N.D. PRIEST[1], Y. KUYANOVA[2], P. POHL[3], M. BURKITBAYEV[2], P.I. MITCHELL[4], L. LEÓN VINTRÓ[4], Y.G. STRILCHUK[5], S.N. LUKASHENKO[5]
[1] School of Health and Social Sciences, Middlesex University, Queensway, Enfield EN3 4SA, UK
[2] Department of Inorganic Chemistry, Al-Farabi Kazakh National University, Almaty, Kazakhstan Republic
[3] Geoecology Department, University of Potsdam, Potsdam, Germany
[4] UCD School of Physics, University College Dublin, Belfield, Dublin 4, Ireland
[5] Institute of Radiation Safety and Ecology, National Nuclear Centre, Kurchatov, Kazakhstan Republic

Abstract: Analysis of strontium-90 in soils showed that most of the SEMIRAD1 and SEMIRAD2 project areas were little contaminated with this radionuclide indicating that the extensive testing of nuclear devices at the STS (including more than 100 groundlevel, aerial and crater-producing explosions) resulted in little dispersed local contamination by fission products, including strontium-90. However, local strontium-90 contamination produced by the Telkem, crater-producing explosions within the SEMIRAD1 study area was evident at distances less that about 3 km from the explosion sites. Within the craters soil strontium-90 concentrations reached 1 kBq kg^{-1}. Around the craters strontium-90 was more widely dispersed than fuel-associated radionuclides and evidence exists to suggest that it is much more mobile within the environment. Within the SEMIRAD1 study area strontium-90 levels were also elevated below the path of the fallout plume produced by the testing of the Soviet Union's first H-bomb in 1953. Radiation doses to residents of the SEMIRAD1 study area were calculated using a modified ECOSYS model. These indicated that strontium-90 was a major contributor to dose in the more contaminated regions around Telkem and close to the village of Sarzhal. Annual doses to adult males living close to Telkem were assessed to currently be about 7 mSv, but these were predicted to fall in line with the physical half-life (28.64 years) of strontium-

90. Localised hotspots that are significantly contaminated with strontium-90 are to be found within Technical Area 4a, which straddles the southern boundary of the SEMIRAD2 study area. About 35 areas of contamination produced by the testing of radiation dispersion devices have been identified. They are considered hazardous since they are unmarked and have been grazed by animals from local farms. They may also present a terrorism hazard. Calculations made using the ICRP biokinetic model for strontium suggest that, under conditions of chronic ingestion, radiation doses to farmers and their families could reach about 4 mSvy^{-1}.

Keywords: strontium-90, Semipalatinsk test site, radiation doses, farmers

1. Introduction

Compared with the levels of some other radionuclides, e.g., caesium-137 and plutonium isotopes, the levels of strontium-90 in soils and biota on the former Soviet Semipalatinsk nuclear test site (STS) have been poorly studied [1] – mostly reflecting difficulties in its measurement. This is because strontium-90 is a β-emitter that produces no photon that can be easily detected using gamma-spectrometry. Its detection, therefore, requires the use of radiochemical procedures that are both complex and time consuming. Nevertheless, strontium-90 is an important radionuclide; it is produced (Fig. 1) in nuclear explosions in quantities that are of the same order as those of caesium-137 (both in terms of mass and radioactivity (3.9 PBq Mt^{-1} for strontium-90 and 5.89 PBq Mt^{-1} for caesium-137 [1])) and under some conditions sufficient amounts may be present in soils and food items grown/produced on these soils to give rise to significant radiation exposures to exposed populations.

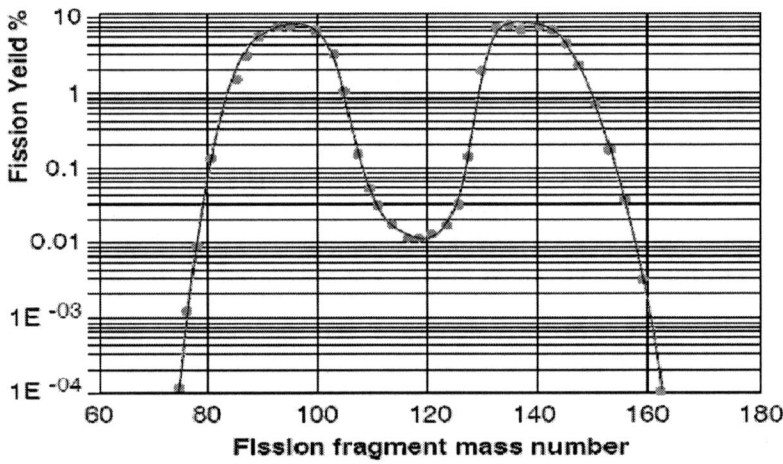

Figure 1. Fission product mass yield spectrum for uranium-235. About 6% of the fission yield is at mass 90. A similar fraction is produced by the fission of plutonium-239

The potential toxicity of strontium-90, as produced by nuclear detonations, results both from the chemical and radiological properties of the radionuclide itself and from the characteristics of the plutonium/uranium fission process that produced it. Strontium is a Group II alkaline earth element and has many characteristics in common with calcium – which is one of the body's major elements. Calcium ions are essential for muscle and nerve function and are accumulated, as the mineral calcium hydroxyapatite, within the skeleton. Moreover, many metabolic pathways exist that ensure that suf-ficient calcium is absorbed from the gut to meet the needs of the body and other pathways exist to maintain and regulate calcium levels within body fluids and tissues – including the skeleton. The chemical properties of strontium ions (Sr^{2+}) are so similar to those of calcium that such ions may ride "piggy-back" on the metabolic pathways that exist for calcium. Accordingly, the fractional uptake of strontium from ingested food in the gut is high (ranging from about 30–60% [2]) and it avidly deposits on the surfaces of bone within the skeleton [3]. Within a short time after intake most of the absorbed strontium is present in the skeleton, with little in soft tissues. Subsequently, strontium may be lost either from the bone surfaces by back ion-exchange into tissue fluids or as a result of the removal of bone mineral during bone resorption or become buried within the bulk of the bone matrix. It follows from the above that when strontium-90 is present in the body it selectively irradiates skeletal tissues. Strontium-90 decays with a physical half-life of 28.64 years to zirconium-90 via yittrium-90. The half life of yittrium-90 (64.1 h) is much shorter than that of strontium-90 and in the skeleton is in equilibrium with its parent. It follows that each strontium-90 soft, short-range β-decay (540 keV) is followed by a hard yittrium-90 β-decay (2.27 MeV) with a very much longer track length in tissues. Together these β-particles irradiate both tissues close to bone surfaces, leading to the possibility of bone tumours, and cells deep within the bone marrow, which are the target cells for leukaemia. Both of these tumour types have been produced by strontium-90 in experimental animals and this radionuclide (along with other internally deposited β-emitting radionuclides) is classified by the World Health Organization's International Agency for Research on Cancer as a Group 1 human carcinogen [4].

The potential toxicity of strontium-90 produced by nuclear explosions is also influenced by the fission process itself. This is because the primary fission yield of strontium-90 is only 0.77% and most of the total strontium-90 produced (equivalent to a total fission yield of 5.77%, Fig. 2) is produced by the decay of other mass 90 nuclides. Of these the most important is krypton-90 which is an isotope of a noble gas with a half-life of 32.3 sec. This decays to strontium-90 via rubidium-90. Therefore, there is a time delay in the production of the largest fraction of strontium-90 during which the parent radionuclide is present as a gas. During this time period the gas cloud above the explosion will have significantly expanded, many larger particles in the cloud would have fallen to earth and the cloud itself would have cooled significantly. It follows that local strontium-90 fallout (and caesium-137 produced from xenon-137) from a surface nuclear explosion will be distributed more widely than many other radionuclides including those that are fuel-associated (e.g., plutonium isotopes and americium-241). Moreover, the size of the particles with which it will be associated will be smaller and the strontium-90 attached to these will be adherent to surfaces and less strongly bound than in similarly sized fused particles produced at high temperature. All of these factors will facilitate strontium-90 migration in soils and its bioavailability by inhalation and ingestion.

Figure 2. Outline of the STS boundary indicating the location of the SEMIRAD1 and SEMIRAD2 study areas

Because of it potential toxicity strontium-90 contamination levels were measured in all soil samples collected for the NATO Science for Peace-funded SEMIRAD1 and SEMIRAD2 projects. Subsequently, these concentrations were used to model the ecosystem transfer of strontium-90 to man and to estimate radiation doses to populations living either on or close to the study areas.

2. Methodology

2.1. DESCRIPTION OF SEMIRAD STUDY SITES

The locations of the SEMIRAD1 and SEMIRAD2 study sites relative to the boundaries of the STS are shown in Fig. 2.

The SEMIRAD1 site (approximately 450 km^2) is a wedge-shaped area of steppe grassland located in the southeast of the STS. It is bounded by the Degelen Mountains to the northwest and by low-lying hills to the north and south. It is relatively flat and is an area traditionally used to graze stock (mostly horses) and for hay collection. The area was chosen for study because it is located close to the village of Sarzhal (population 2000) and because it is subject to contamination by run-off waters from the Degelen Mountains (the site of 239 underground, tunnel nuclear tests), by fallout from two small nuclear tests (Telkem1 and Telkem2) which produced craters that subsequently filled with water, and because the southeast of the area was intercepted by the fallout plume produced by the Soviet Union's first thermonuclear explosion (12th August 1953) at Ground Zero – located south of the SEMIRAD2 area. A small number of livestock farms are also located in the eastern region of the site.

The SEMIRAD2 site is larger than the above (approximately 800 km^2). It was chosen for evaluation because it is located to the north of the ground zero technical area and it may have become contaminated by the 26 ground and 87 atmospheric tests at this site. It contains several saltpans and is bisected by a small seasonal stream – the Ashyozek River. Unexploited mineral resources – including gold and manganese – are thought to be present. The SEMIRAD2 site contains farms that are managed by local Kazakh families on behalf of owners in Pavlodar. Crops traditionally grown in the area include millet and watermelons. The land was attractive for study because it is close to the city of Kurchatov, the local village of Maisk, and is bounded at its northern edge by the STS site boundary (which would make it easy to release for unrestricted use – if found to be uncontaminated). Finally, it is close to/includes sites of suspected Soviet nuclear experiments – including the northern part of Technical Area 4a where ground-based, aircraft-launched and artillery-fired radiation dispersal devices (RDDs), filled with strontium-90 and other fission products, were tested in the 1950s [5].

2.2. FIELD SAMPLING

At both the SEMIRAD1 and SEMIRAD2 sites, evaluations of the radiological conditions across the general site have been undertaken using a 2 × 2 km sampling grid. At the former site the grid employed was irregular [6] but, in order to conform more closely to IAEA monitoring recommendations, a regular grid was used for sample collection during SEMIRAD2. In both cases sampling locations were located by GPS. At each sampling site gamma-monitoring was undertaken and soil samples were collected for radionuclide analysis. At all sites soil samples (20 × 20 × 10 cm deep) were collected after the removal of surface vegetation – where present. In addition, at each tenth sampling site a short soil profile sample was collected with sampling at 0–5 cm, 5–10 cm and 10–15 cm depth. The soil samples were analysed by gamma-spectrometry, mostly at the NNC – Institute of Radiation Safety and Ecology (IRSE), to determine their natural radionuclide content plus their content of americium-241 and caesium-137.
In addition, plutonium isotopes and strontium-90 were measured. Usually these were separated from samples using radiochemical techniques (mostly at the NNC – Institute of Nuclear Physics (INP) and/or al-Farabi Kazakh National University) and analysed by

Figure 3. The Telkem2 crater produced on 12 November 1968 by the simultaneous explosion of three 240 t fission devices

Figure 4. Evaluation of a RDD site within Technical Area 4a where significant contamination by strontium-90 is present

α-spectrometry and β-spectrometry, respectively. However, plutonium was sometimes determined by XRF. Surface and well water samples were also collected for radionuclide analysis [7].

In addition, at both study sites, further samples were collected at sites of special interest. For SEMIRAD1 these sites were located around the craters produced by the Telkem1 and Telkem2 civil, excavation experiments of 1968. These craters were produced by low-yield (240 t), plutonium-fuelled fission devices exploded at a depth of about 30 m below the ground surface. At Telkem1 a single device was exploded producing a circular crater. At Telkem2 three devices were exploded simultaneously producing a 300 × 250 m elongated crater. Subsequently, both craters filled with fresh water. In the case of Telkem2 this produced a lake approximately 130 m long, by 45 m wide and 7–10 m in depth (Fig. 3). Sampling of the 16 km^2 area around these craters was undertaken using a 500 × 500 m grid. Close to the craters the sampling locations were closer – 150–50 m separation. At Telkem1 an alternative strategy was also employed with sampling along eight equally spaced, radial and diverging transects centred on the centre of the crater. As for the examination of the general site the soil samples collected at each site were analysed to determine their strontium-90 and other radionuclide content. The radionuclide content of water samples was also determined.

All the soil contamination data collected for the SEMIRAD1 project site were entered into the DECODA GIS database maintained by IRSE. Subsequently they, and data on the levels of radionuclides in water, were used for ecosystem transfer modelling to estimate doses to the human populations living on and close to the SEMIRAD1 study area. This modelling was undertaken in collaboration with Mouchel Parkman plc, a consultancy company leading a STS land utilisation study funded by the United Kingdom Department for International Development.

On the SEMIRAD2 project area the site of special interest studied was Technical Area 4a – a site which had been used for RDD testing. Other potential sites (Technical Areas 8 and 9) were also examined, but these proved not to be contaminated. The examination of Area 4a is still in progress and is being undertaken in association with another project team led by IRSE and RWE-NUKEM plc and funded by the United Kingdom Department of Trade and Industry. Initial studies have shown that this technical area is, in places heavily contaminated by strontium-90 and some other radionuclides. The locations of these deposits are being determined by a walk over survey using surface contamination monitors and hand-held gammadose-rate meters (Fig. 4). As yet no samples have been removed for laboratory analysis – a precaution to prevent laboratory contamination and all evaluations are being conducted *in situ*. About 35 sites, contaminated mostly by strontium-90, have been identified within Technical Area 4a (Figs. 5 and 6).

Given that Area 4a has been open to grazing by animals, particularly horses, from nearby farms it was decided to further investigate strontium-90 levels in soils close to the farms (Fig. 5) where contaminated animals may have deposited faeces in enclosures and close to farm dwellings, in well waters used by the farms, in fodder and in the meat and bones of farm animals. This was undertaken to determine the potential impact of the ingestion of strontium-90-contaminated food and water by the farmers and their families. This method of assessment was chosen, rather than ecosystem modelling, because it is difficult to predict where animals are grazing and the extent of their exposure

STRONTIUM-90 CONTAMINATION WITHIN THE SEMIPALATINSK TEST SITE   93

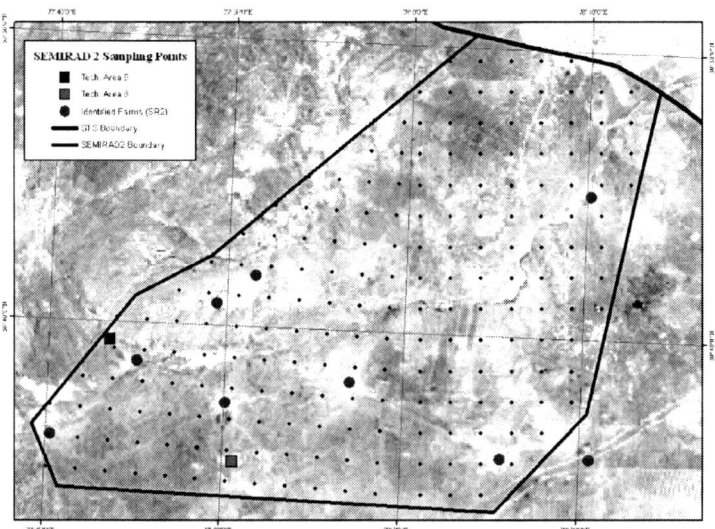

Figure 5. Landsat image of the SEMIRAD2 study area. Part of Technical Area 4a (not shown) straddles the southern portion of the site and is close to several farms

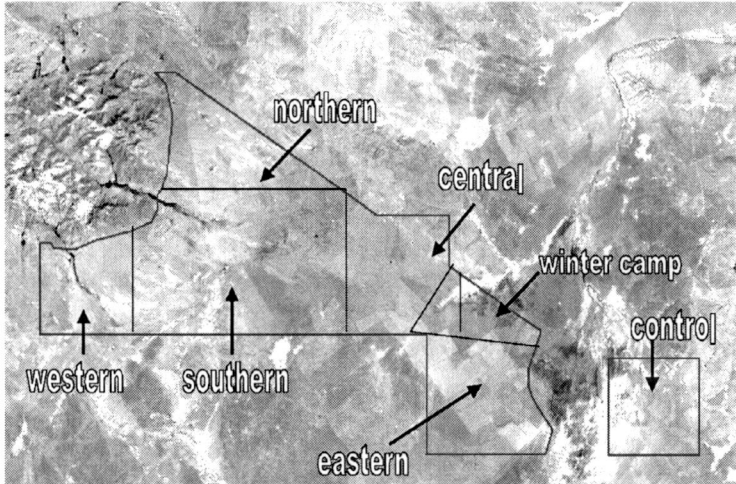

Figure 6. Landsat image of the SEMIRAD1 study area showing the zones used for radionuclide analysis. Also shown is the location of the winter camp area that contains all known farms. For dosimetry the nomenclature was modified. The western, northern and southern zones were combined and re-designated – summer camp 1 area. The central zone was re-designated as the general Telkem area and the eastern zone is referred to as the summer camp 2 area

to contaminated soils. Moreover, if the levels of strontium-90 in food can be measured, and suitable habit data is available to estimate intake, direct estimates of ingestion will provide better estimates of dose.

2.3. MODELS EMPLOYED FOR DOSE ESTIMATION

A modified form of the ECOSYS radioecological model was employed for the estimation of population doses arising from the radioactive contamination of the SEMRAD1 study area. This model was developed by GSF Neuherberg in Germany [8]. ECOSYS assesses the consequences of radioactive releases to the environment and accounts for the transfer of radionuclides through the food chain, the inhalation of radioactive aerosols – produced by the re-suspension of contaminated soils – and doses from external irradiation. ECOSYS considers many transfer pathways that are relevant to both short-term and long-term exposures. For the STS assessment ECOSYS short-term exposure processes and pathways were ignored [9]. The input data used for modelling were the measured levels of strontium-90 and other radionuclides in soils and drinking water, estimated concentrations of dust in air, published transfer factors and habit data for local populations. Some of the habit data used was from published sources, but others were obtained by direct communication with local populations. Radiation doses were calculated for the years 1990 (when the STS was closed) through to 2059. Doses were calculated for farmers and their families (men, women, children and infants) living on the SEMIRAD1 territory in farms located to the southwest of the Telkem explosion sites and to hypothetical farmers living elsewhere within the study area. Both the total dose ($mSvy^{-1}$) and the doses arising from exposures to individual radionuclides – including strontium-90 – were calculated. Full details of the model have been published by Pohl [9].

As yet, no radioecological modelling has been undertaken to estimate doses to farmers living within the SEMIRAD2 study site. Instead, data on the concentration of strontium-90 in food and drinking water are being used to directly estimate doses and the size of accumulated body burdens as a function of exposure time. Doses were calculated using the International Commission on Radiological Protection (ICRP) dosimetric model for strontium [2]. The ICRP strontium model was implemented using an algorithm developed by Birchall [10, 11]. This is encoded within a programme code (RDMS) developed at Middlesex University and its accuracy has been verified by the comparison of results produced with those published by the ICRP. The model has been run to estimate strontium-90 body burdens and radiation doses to skeletal tissues following chronic intakes of strontium-90-contaminated food and water.

3. Results

3.1. SEMIRAD1 STUDY AREA

For convenience of description the SEMIRAD1 study area was divided into zones (Fig. 6). For each of these the median strontium-90 content per unit mass of soil was calculated ($Bq\ kg^{-1}$) using the results obtained for the general survey sites. The results are presented in Fig. 7. This shows that in the northern and southern zones, which were beyond the trajectory of the contamination plume from the August 1953 thermonuclear test at ground zero, the median concentration of strontium-90 in the top 10 cm of soil is below 10 $Bq\ kg^{-1}$ – equivalent to about twice the global fallout average for the northern

STRONTIUM-90 CONTAMINATION WITHIN THE SEMIPALATINSK TEST SITE

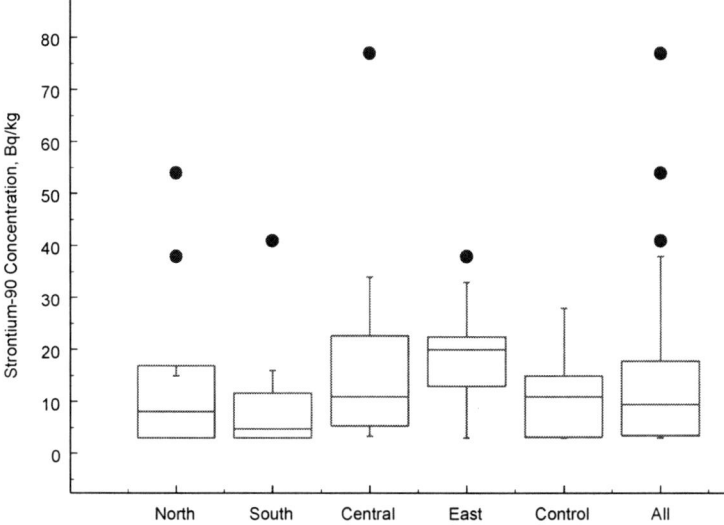

Figure 7. Box and whisker plots showing the measured concentrations of strontium-90 in soil samples collected on the SEMIRAD1 site during the general site survey

hemisphere. Towards the east the concentrations measured were about two times higher demonstrating the influence of the 1953 explosion. The results are consistent with caesium-137 maps [1], which indicate that the plume from the thermonuclear explosion clipped the north of our central zone, close to Telkem, then passed over the centre of the east zone and intercepted the south of the control zone, close to Sarzhal village. In all zones, the strontium-90 concentration distributions determined were skewed towards low concentrations – consistent with contamination by a pollutant that is present with a patchy distribution (i.e., with hotspots).

One unexpected finding for the general site was the failure of the explosions at Telkem to significantly impact upon the strontium soil concentrations measured in the central region. This is an artefact and is a consequence of the ~2 km separation between sampling points for the general survey. The much more detailed survey of the Telkem general area and the Telkem1 and Telkem2 craters showed a contaminated area, with soil strontium-90 concentrations exceeding 10 Bq kg^{-1} that extended out to about 3 km from the explosion site (see SEMIRAD1 report for more details [6]). Around Telkem1 strontium-90 levels were much lower than in the area around Telkem2.

In the craters and within the crater walls strontium-90 levels were considerably elevated above those found across the general site. At Telkem2 the highest median concentrations, about 350 Bq kg^{-1}, were found in the lake sediments and at the edge of the lake (Fig. 8). However, contamination within the crater itself was not evenly distributed and samples collected in the north wall of the crater were much more contaminated than either those in the south wall or those in the lake sediments. The highest measured concentration of strontium-90 exceeded 1 kBq kg^{-1}. Within the bulk of the crater the median concentrations were lower (about 200 Bq kg^{-1}) than at the lake edge and close to

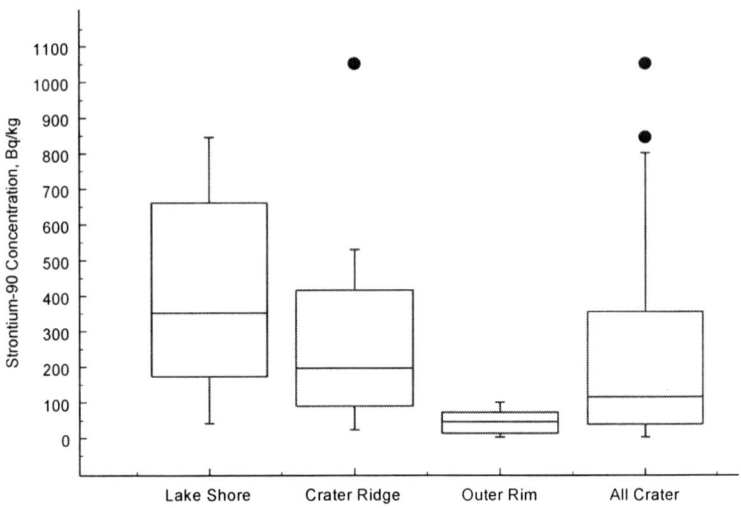

Figure 8. Measured concentrations of Strontium-90 in soil samples collected within the Telkem2 crater

the outer edge of the crater typical concentrations of strontium-90 were lower again at about 50 Bq kg^{-1}. Similar concentrations were measured in soil samples collected at the smaller Telkem1 crater. At this site most contamination lay to the east and southeast of the crater. Within the crater walls only about 20% of the strontium-90 was associated with smaller soil particles – less than 0.5 mm in diameter. In contrast, at distances greater than about 150 m from the point of explosion most, about 80%, of the radionuclide was associated with smaller particles. This result is consistent with expectations based on the delayed production of strontium-90 following fission. Similarly, strontium-90 was much more widely dispersed than plutonium isotopes (and americium-241) – the concentrations of which fell by 4 orders of magnitude within 200 m of the explosion point. Consideration of caesium-137: strontium-90 ratios as a function of distance from the centre of the Telkem2 crater suggests that strontium-90 may be more mobile in the environment than caesium-137 [6]. As a consequence a poor correlation ($r^2 = 0.47$) was found between strontium-90 and caesium-137 concentrations in SEMIRAD1 soils. In contrast, caesium-137 levels were better correlated with plutonium-239 concentrations ($r^2 = 0.79$).

Because of their skewed distributions the concentrations of strontium-90 inputted into the ECOSYS model were the geometric mean concentrations for soil samples collected in the summer camp 1 area, in the wider Telkem area, in the winter camp area and in the summer camp 2 area (see Fig. 9). These were used first to calculate the current concentrations of strontium-90 (and caesium-137, hydrogen-3, americium-241, plutonium-239+240) in food and then in the human population that consumed the food. In all areas the predicted concentrations of plutonium and americium in food were vanishingly small (<1 mBq kg^{-1}). In all areas except the Telkem area the predicted concentrations of strontium-90 in food were lower than those of caesium-137. Typical values calculated for animal products are presented in Table I. The calculations indicate that the highest concentrations of both radionuclides will be present in horse meat and the most significant source of both will be water within the Telkem craters. These levels are somewhat higher

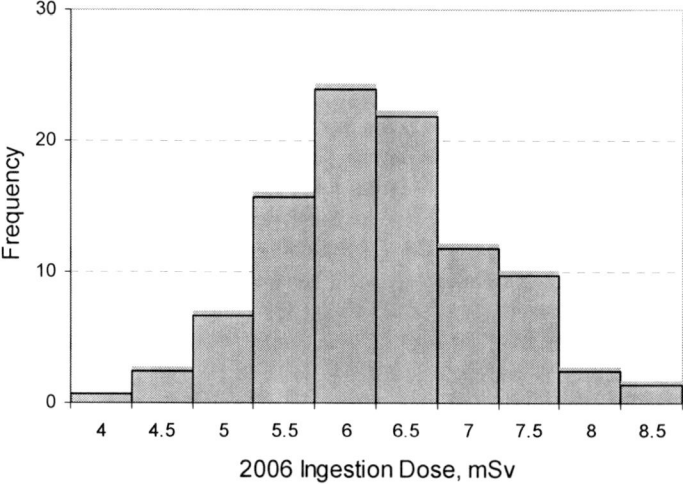

Figure 9. The calculated probability distribution for the calculated total ingestion dose to men in the Telkem area, 2006

TABLE I. Concentrations (Bq kg^{-1}) of caesium-137 and strontium-90 in some animal products, as predicted using the ECOSYS model for SEMIRAD1

Area	Product	Cs-137	Sr-90
Summer camp 1	Horse milk	11.3	2.5
Summer camp 2	Horse milk	30.4	18.1
Winter camp	Horse milk	29.2	20.4
Telkem area	Horse milk	71.3	161
Summer camp 1	Mutton	4.1	0.5
Summer camp 2	Mutton	11.1	3.3
Winter camp	Mutton	10.6	3.8
Telkem area	Mutton	25.9	29.5
Summer camp 1	Cows milk	1.4	0.3
Summer camp 2	Cows milk	3.8	2.4
Winter camp	Cows milk	3.6	2.7
Telkem area	Cows milk	8.9	21.4
Summer camp 1	Horse meat	71.6	7.1
Summer camp 2	Horse meat	192	51.8
Winter camp	Horse meat	185	58.3
Telkem area	Horse meat	451	459

than those measured in food collected within Sarzhall village: potato, 24.2 ± 5.3 Bq kg^{-1}; curds, 0.39 ± 0.14 Bq kg^{-1}; cows milk, 0.47 ± 0.19 Bq kg^{-1}; unspecified meat 1.7 ± 0.4 Bq kg^{-1}. However, the well water in this village contained much less strontium-90 than that collected on the SEMIRAD1 site.

For men, women and children the consumption of contaminated water was also calculated to be the most important exposure pathway. Infants that were breast fed ingested much less strontium-90 than those that drank water, eat meat, etc. It follows that the calculated annual doses in 2001 to infants in the summer and winter camp areas were always insignificant, less than about 30 μSvy^{-1}, and dominated by doses resulting from the accidental ingestion (pica) of plutonium isotopes. Within the Telkem area the calculated doses to infants for this year were higher (170 μSvy^{-1}), in line with higher plutonium soil contents, but still of little radiological significance. In contrast, the calculated dose from the ingestion of strontium-90 was below 1 μSvy^{-1}. Table II shows the calculated total ingestion dose and the ingestion dose from strontium-90 for men, women and children during 2001. It can be seen that in the summer camp 1 area strontium-90 contributes little to total dose. In this area the major source of dose is tritium – consequent to the ingestion of stream waters running out of the Degelen Mountains [12]. In contrast, in other areas strontium-90 gives rise to the largest calculated doses and is the major contributor to total dose. Only for the Telkem area are the 2001 doses calculated to always exceed 1 mSvy^{-1}. The highest calculated annual ingestion dose is for men (8.45 mSv) and of this 92% is calculated to arise from the ingestion of strontium-90. This result emphasises the importance of strontium-90 as a contributor to dose in areas contaminated by this radionuclide.

TABLE II. Calculated doses (mSvy^{-1}) from ingestion in 2001

Area	Age group	Sr-90	Total
Summer camp 1	Men	0.05	1.25
Summer camp 2	Men	0.77	1.04
Winter camp	Men	0.81	1.07
Telkem area	Men	7.79	8.45
Summer camp 1	Women	0.05	1.20
Summer camp 2	Women	0.71	0.95
Winter camp	Women	0.74	0.97
Telkem area	Women	7.27	7.86
Summer camp 1	Children	0.05	0.92
Summer camp 2	Children	0.61	0.78
Winter camp	Children	0.65	0.81
Telkem area	Children	6.12	6.54

Given the above, as expected, the calculated doses for other years is a function of the physical half life of strontium-90. The corresponding annual doses calculated for 1990, 2006 and 2059 are 10.95, 7.51 and 2.18 mSv, respectively. In addition to ingestion doses the doses from inhalation and external exposures were also calculated, but again these contributed little to total dose. For example, in the summer and winter camp areas the calculated 2006 inhalation doses ranged from 0.14 to 1.15 µSv and the corres-ponding external doses ranged from 5 to 60 µSv [9].

It is axiomatic that even if the modelled pathways are secure, all modelled assessments are subject to significant uncertainties that arise because of the use of non-site-specific transfer factors during modelling. It follows that in order to estimate the likely impact of these uncertainties on total ingestion dose, an analysis was undertaken using modified model parameters. For the uncertainty analysis the modified ECOSYS model was run with sets of 100 values for each parameter. These values were randomly allocated either within their known probability distributions or within a reasonable range. A histogram produced in this way and showing the calculated range of ingestion doses for men in the Telkem region in 2006 is reproduced in Fig. 9. The level of uncertainty in the central estimates of dose indicated is considered reasonable given the complexity of the model. Additional, unestimated uncertainty will result from uncertainties in the human dosimetric model for strontium.

3.2. SEMIRAD2 STUDY AREA

As yet, analysis of the general site survey samples collected within the SEMIRAD2 project area has not been completed. However, those analyses completed, to date, suggest that this site is little contaminated by strontium-90 over most of its area. The results of the 20 completed soil analyses are compatible with the strontium-90 soil concentrations measured at SEMIRAD1 – with a median concentration of 9.6 Bq kg^{-1} and with corresponding lower and upper 95% confidence intervals of 2.1 and 42.9 Bq kg^{-1}. It follows that, should ECOSYS modelling be applied, the annual radiation doses from strontium-90 calculated for farmers resident within the SEMIRAD2 study area would be similar to those calculated for farmers using the summer camp areas of the SEMIRAD1 site (<1 mSvy^{-1}).

In contrast to the above, Technical Area 4a, which straddles the southern SEMIRAD2 boundary is highly contaminated by strontium-90 (plus other long-lived fission products at much lower concentrations) at 35 locations that were identified/ confirmed by the walkover survey conducted in association with the UK DTI project team (Fig. 10). Indications are that, at these sites, strontium-90 surface concentrations may sometimes be as great as 500 MBq kg^{-1} of soil and that significant concentration of the radionuclide may be present down to 10 cm below the soil surface. For example, the concentrations of strontium-90 in one profile from a moderately contaminated soil (to the side of a main deposit) showed that the concentrations decreased as follows: 2337 ± 32 Bq kg^{-1} (0–5 cm); 271 ± 6.4 Bq kg^{-1} (5–10 cm); 81.7 ± 4.3 Bq kg^{-1} (10–15 cm); 81.2 ± 8.1 Bq kg^{-1} (15–20 cm). The indicated peak concentrations at the 35 locations are 500 times higher than those found in the Telkem craters and present a significant hazard to local farmers – with the degree of risk depending upon occupancy factors for either the farmers or their grazing stock. They may also present a terrorism hazard – since the radionuclide present at

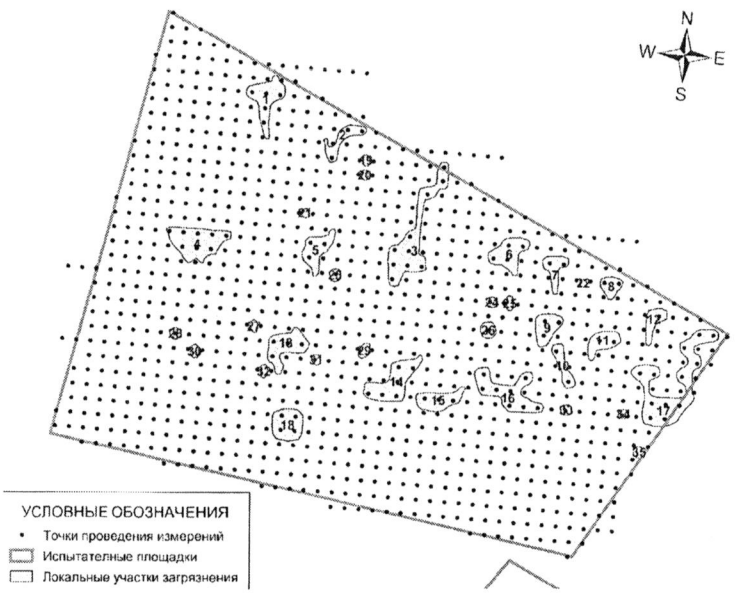

Figure 10. Distribution of sites of local contamination within Technical Area 4a. The locations of 1,059 survey points used for the walk-over assessment are indicated

high concentrations could be easily harvested for the manufacture of radiological weapons [5].

The walkover survey also revealed the presence of animal droppings and animal bones that clearly indicated that the pastures within Technical Area 4a had been grazed – at least by horses from at least one local farm – Tulpar Farm. In 2004, herds of up to 700 horses could be found grazing in the area traditionally grazed by this farm. In view of the above the sampling strategy was extended to include the collection of plant materials within Technical Area 4a and fodder and food samples at local farms. In addition, animal bones were collected from the local steppe and at the farms. These are to be used as an indicator of the likely extent of farm animal contamination and, in turn, the likely levels of strontium-90 in meat. Surface contamination monitors clearly demonstrated that some of the bones collected contained β-emitting radionuclides. The results for samples collected in 2005 are shown in Table III. It can be seen that the plant materials collected near a contamination hotspot in Technical area 4a were substantially contaminated with strontium-90. Similarly, this radionuclide was present in horse droppings, bones and meat collected either in Area 4a or at the adjacent Tulpar Farm. This observation is consistent with the transfer of strontium-90 from food plants to grazing animals. In contrast, milk and milk products collected at the farm contained only low concentrations of the radionuclide. As yet other samples, including well waters, have not been analysed for strontium-90.

TABLE III. Concentration of strontium-90 in plant materials and food collected during 2005. All samples were collected either from Technical Area (TA) 4a or from farms close to this area

Sample	Collection site	Sr-90 Bq kg^{-1}
Absinth (washed)	TA 4a	28714 ± 268
Steppe feather grass (washed)	TA 4a	3056 ± 573
Ephedera berries (washed)	TA 4a	4835 ± 357
Horse droppings	TA 4a	880 ± 103
Sheep bones (rib)	TA 4a	3557 ± 388
Horse bone (skull)	TA 4a	3215 ± 377
Meat (sheep)	Tulpar farm	368 ± 27
Curds	Tulpar farm	0.39 ± 0.08
Cow's milk	Tortkuduk farm	0.39 ± 0.08
Horse's milk	Tortkuduk farm	0.41 ± 0.02
Sour cream	Tortkuduk farm	0.35 ± 0.08
Butter	Tortkuduk farm	0.46 ± 0.01

The farmers eat meat that they produce (under conditions regulated by the farm owners – typically one ~50 kg sheep per month per family, and/or horse meat). It follows that the demonstrated presence of strontium-90 in animal bones and the meat sample measured indicate that the farm families may ingest radiologically significant amounts of strontium-90 each day. Pending the results of further strontium-90 analyses, scoping studies have been undertaken to estimate the magnitude of putative radiation doses, from strontium-90, to adult farmers assuming the daily consumption of contaminated meat. It follows, that dosimetry calculations, made using the ICRP biokinetic model for strontium, have assumed a nominal daily ingestion rate of 1 Bq day^{-1} of strontium-90 and a fractional absorption of this radionuclide from the gut of 0.4. The model was used to predict the amounts of radiostrontium accumulated by tissues and organs over a period of 50 years (18,263 days) of continuous intake. Subsequently, radiation doses per unit intake were calculated as a function of time. Finally these doses were used to estimate the radiation doses received by farmers assuming the daily consumption of 0.25 kg of meat (30 kg meat, from 50 kg carcass, eaten by a family of four persons over a period of 30 days). The contamination level in the meat collected at Tulpar Farm (368 Bq kg^{-1}, cf., 8 mBq kg^{-1} for plutonium-239+240) was used to derive this estimate (92 Bq day^{-1}). It should be noted that 368 Bq kg^{-1} is almost four times higher than the generic action level for strontium-90 in foodstuffs (0.1 kBq kg^{-1}) as established by the joint FAO/WHO Codex Alimentarius Commission (CAC) [13]. However, 368 Bq kg^{-1} is lower than the European Union maximum permitted level of strontium isotopes in foodstuffs following a nuclear accident/radiological emergency [14]. This limit is 750 Bq kg^{-1} for foodstuffs other than infant foods and dairy products.

The predicted body and tissue burdens, as a function of time from the start of chronic daily ingestion of contaminated food, are shown in Fig. 11. This shows that for a daily intake of 1 Bq the predicted body burden rises from a calculated 0.3 Bq at 1 day of exposure, through 10 Bq and 54 Bq after 100 and 1,000 days, respectively, to 235 Bq after 50 years of chronic exposure. Beyond the first few days most of this was present in the skeleton. For example, at 50 years all but 4 Bq (1.7%) are predicted to be in bone compartments. Table IV shows the corresponding predicted concentrations of strontium-90 in bone compartments – this information is required for dosimetry. It can be seen that although for modelling it is assumed that strontium is initially deposited on bone surfaces

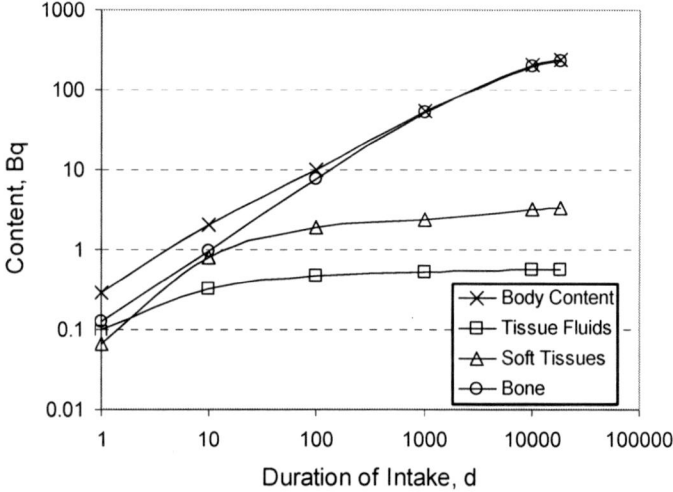

Figure 11. Predictions of the ICRP biokinetic model for Sr-90 assuming the chronic ingestion of 1 Bq day^{-1} for 50 years

TABLE IV. Predicted strontium-90 content of bone compartments defined by the ICRP biokinetic model for strontium-90 – assuming a daily ingestion rate of 1 Bq day^{-1}

Model compartment	Duration of exposure (days)					
	1	10	100	1000	10000	18263
Cortical bone volume (non-exchangeable)	0.000	0.003	0.57	18.7	129.7	159.8
Cortical bone volume (exchangeable)	0.005	0.19	2.57	5.520	5.92	5.96
Cortical bone surface	0.050	0.23	0.35	0.394	0.42	0.43
Trabecular bone volume (non-exchangeable)	0.000	0.004	0.70	19.7	56.18	56.98
Trabecular bone volume (exchangeable)	0.006	0.24	3.20	6.88	7.37	7.42
Trabecular bone surfaces	0.062	0.29	0.43	0.49	0.53	0.53

most of this radionuclide is transferred to bone volume compartments – particularly the cortical bone volume compartments that have a low rate of turnover.

The published ICRP Ingestion Dose Coefficient (Sv Bq^{-1} to Age 70 years) for strontium-90 is 2.8×10^{-8} Sv Bq^{-1} [2]. During 50 years the amount of radionuclide ingested, assuming a 92 Bq day^{-1} ingestion rate, is ~1.7 MBq, which gives an effective dose of 47 mSv (or an average of about 1 $mSvy^{-1}$). However, this calculation is inappropriate for a uniform chronic dose to adults from age 20 to age 70 since the mean period from intake to age 70 is only 25 years. Preliminary calculations made using the ICRP model suggest that over the period of 50 years the strontium-90 skeletal content will rise from 0 Bq at time zero to 21 kBq after 50 years of chronic intake. The cor-responding doses to bone surfaces are 0 and ~62 $mSvy^{-1}$. The latter dose corresponds to an equivalent whole body dose (using a tissue weighting factor of 0.01 for bone surfaces and taking account of dose to other tissues) of about 4 mSv. It follows that, at a contamination level of 368 Bq kg^{-1}, the strontium-90 in meat is unlikely to give rise to a radiation dose large enough to be of serious concern – although the predicted doses after a few thousand days of chronic intake do exceed the 1 $mSvy^{-1}$ dose limit for members of the public in Kazakhstan.

4. Conclusions

Given the above information it can be concluded that: Most of the SEMIRAD1 and SEMIRAD2 project areas are little contaminated with strontium-90 above levels that greatly exceed those expected in the northern hemisphere from global weapons fallout. An average finding of <10 Bq kg^{-1} is consistent with the conclusions of a previous IAEA investigation team [1], which reported concentrations of <20 Bq kg^{-1} close to Sarzhal and <19 Bq kg^{-1} at the Beriozka State Farm. The results indicate that the extensive testing of nuclear devices at the STS (including more than 100 ground-level, aerial and crater-producing explosions that were not contained) resulted in little dispersed local contamination by fission products, including strontium-90.

Local strontium-90 contamination produced by the Telkem, crater-producing explosions within the SEMIRAD1 study area was restricted to within about 3 km from the explosion point at Telkem2 and within a much shorter distance at Telkem1. Moreover, the levels of strontium-90 in soils were only significantly elevated within the crater walls and in the surface layers (0–5 cm) of soil within a few hundred metres of the crater. Around the craters strontium-90 was more widely dispersed than fuel-associated radionuclides and evidence exists to suggest that it is much more mobile within the environment.

Within the SEMIRAD1 study area strontium-90 levels were also elevated below the path of the fallout plume produced by the testing of the Soviet Union's first H-bomb at Ground Zero (12th August 1953).

Radiation doses to residents of the SEMIRAD1 study area were calculated using a modified ECOSYS model. The doses calculated indicated that strontium-90 was a minor contributor to dose in the west of the area close to the Degelen Mountains, but contributed most of the dose in the more contaminated regions around Telkem and close to the village of Sarzhal. Most of the predicted doses resulted from the ingestion of water and were below 1 $mSvy^{-1}$ – in agreement with similar doses that have been calculated by Semiochkina and her colleagues [15] for residents of farms near Ground Zero and close to the Balapan atomic lake. However, the calculated current annual dose, from strontium-90, to adult males living close to Telkem was about 7 mSv. This dose is not dissimilar to the 10 mSvy^{-1} dose calculated by IAEA for the future permanent settlement of the ground zero area [1]. Since the 7 mSvy^{-1} dose was predicted to fall in future years, in line with the physical half-life of strontium-90, it is unlikely that radiation doses in the Telkem region will exceed the IAEA Basic Safety Standard (BSS) for relocation and permanent resettlement – 30 mSv month^{-1} for relocation and a life-time dose of >1 Sv for permanent resettlement [1]. However, the IAEA BSS relates to total dose (natural plus anthropogenic) and the acceptability of the 7 mSv annual dose from strontium-90 (8.45 mSv from all anthropogenic sources) will depend upon local, natural, ambient background radiation levels. Even so, for reasons of comparison, it is worth pointing out that 8.45 mSv is much lower than IAEA estimates of the annual exposures in typical high natural background areas – 17 mSv [1].

Localised hotspots that are significantly contaminated with strontium-90 are to be found within Technical Area 4a, which straddles the southern boundary of the SEMIRAD2 study area. About 35 areas of contamination have been identified and, in some, strontium-90 concentrations in soils may reach 500 MBq kg^{-1}. These hotspots were produced by the testing of radiation dispersion devices during the 1950s. The identified deposits are considered hazardous since the contaminated sites are unmarked and have been grazed by horses and sheep from local farms. They may also present a terrorism hazard [5].

The extent of the transfer of soil strontium-90 within Area 4a to farm animals and then to man has yet to be determined with any precision. However, a meat sample collected from the Tulpar Farm contained almost 400 Bq kg^{-1} of strontium-90. The extent to which this level is typical is currently being determined by the analysis of further meat samples and bones collected in the area. This is important because if the levels of strontium-90 in meat are found to routinely exceed the CAC action level, then action made need to be taken to either reduce livestock exposures to the radionuclide or to prevent the sale of the meat produced.

Calculations made using the ICRP biokinetic model for strontium, implemented within the RDMS code developed by Middlesex University suggest that under conditions of chronic ingestion – 250 g of the above meat per day for 50 years – radiation doses to farmers and their families (up to 4 mSvy^{-1}) would exceed the current dose limit for members of the public within the Republic of Kazakhstan (1 mSvy^{-1}) but are unlikely to result in a significant risk of disease or exceed the IAEA BSS [1].

5. Acknowledgements

This work was supported by the NATO Science for Peace programme (SEMIRAD Project, Contract SfP-976046(99) and SEMIRAD 2 Project, Contract SfP-980906(04)). We also wish to thank colleagues at RWE-NUKEM plc who shared information on the radiological condition of Technical Area 4a and at Mouchel Parkman plc who helped support the ECOSYS modelling undertaken by Patricia Pohl.

References

1. IAEA (1998) Radiological conditions at the Semipalatinsk test site, Kazakhstan: preliminary assessment and recommendations for further study. *Radiological Assessment Report Series*. International Atomic Energy Agency, Vienna.
2. ICRP (1993) Age-dependent doses to members of the public from intake of radionuclides: part 2 ingestion dose coefficients, ICRP Publication 67, *Annals of the ICRP*, **23**.
3. Priest, N. D. (1990) The distribution and behaviour of metals in the skeleton and body: studies with bone-seeking radionuclides. In: *Trace Metals and Fluoride in Bones and Teeth*. Eds: Priest, N. D. and Van de Vyver, F. CRC, Boca Raton, FL, 83–140.
4. IARC (2001) Ionizing radiation, part 2: some internally deposited radionuclides. *IARC Monographs on the Evaluation of Carcinogenic Risk to Humans*, **78**, World Health Organization International Agency for Research on Cancer. IARC, Lyon, 478–481.
5. Logachev, B. (2002) Radioecological consequences of radiation dispersion device testing in the Semipalatinsk test site (trans.). *Bull. Atomic Energy*, **12**, 62–67 (Russian).
6. Priest, N. D., Burkitbayev, M., Artemyev, O., Lukashenko, S. and Mitchell, P. (2003) Investigation of the radiological situation in the Sarzhal region of the Semipalatinsk nuclear test site. Final Report of *NATO Science for Peace Project* SfP-976046(99). Available from Middlesex University website: www.mdx.ac.uk/risk/staff/docs/nick16.pdf . Last accessed September 2006.
7. León Vintró, L., Mitchell, P. I., Omarova, A., Burkitbayev, M., Jimenez Napoles, H. and Priest, N. D. (2006) Americium, plutonium and uranium contamination and speciation in well waters, streams and atomic lakes in the Sarzhal Region of the Semipalatinsk nuclear test site, Kazakhstan. *J. Environ. Radioact.* (Special Edition) in press.
8. Müller, H. and Pröhl, G. (1993) ECOSYS-87: a dynamic model for assessing radiological consequences of nuclear accidents. *Health Phys.*, **64**, 232–252.
9. Pohl, P. I. (2005) Modelling human exposures to anthropogenic radionuclides in the environment of the Sarzhal region of the former nuclear weapons test site Semipalatinsk in Kazakhstan. *Diploma thesis*, Universität Potsdam, Germany.
10. Birchall, A. (1990) Appendix to Priest, N. D. (1990) Sensitivity testing of an age-related, multicompartment dosimetric model for bone seeking radionuclides in man. *Health Phys.*, **57**(suppl.), 229–242.
11. Priest, N. D. (1989) Alpha-emitters in the skeleton: a personal evaluation of the risk of leukaemia following intakes of plutonium-239. *Br. Inst. Radiol. (BIR) Report* **21**, 159–166.
12. Mitchell, P. I., León Vintró, L., Omarova, A., Burkitbayev, M., Jimenez Napoles, H. and Priest, N. D. (2005) Tritium in well waters, streams and atomic lakes in the EAST Kazakhstan Oblast of the Semipalatinsk Nuclear Test Site. *J. Radiolog. Prot.*, **25**, 141–148.
13. CAC (1991) Guideline levels for radionuclides in foods following accidental nuclear contamination. Codex Alimentarius, General Requirements, Section 6.1. FAO/WHO Food Standards Programme, Rome.
14. European Union (1989) Council Regulation 89/2218/EURATOM, 18th July.
15. Semiochkina, N., Voigt, G., Mukusheva, M., Bruk, G., Travnikova, I. and Strand, P. (2004) Assessment of the current internal dose due to ^{137}Cs and ^{90}Sr for people living within the Semipalatinsk test site, Kazakhstan. *Health Phys.*, **86**, 187–192.

CHAPTER 10

THE JOINT CONVENTION ON THE SAFETY OF SPENT FUEL MANAGEMENT AND ON THE SAFETY OF RADIOACTIVE WASTE MANAGEMENT: AN INSTRUMENT TO ACHIEVE A GLOBAL SAFETY

P. RISOLUTI
International Atomic Energy Agency, Wagramer Strasse 5, P.O. Box 100, A-1400 Vienna Austria

Abstract: The Joint Convention on the Safety of Spent Fuel Management and the Safety of Radioactive Waste Management (the Joint Convention) is the first legally binding international treaty in the area of radioactive material management. It was adopted by a Diplomatic Conference in September 1997 and opened for signature on 29 September 1997. The Convention entered into force on 18 June 1998, and to date (May 2006) has been ratified by 41 countries. The Joint Convention applies to spent fuel and radioactive waste resulting from civilian application. Its principal aim is to achieve and maintain a high degree of safety in their management worldwide. The Convention is an incentive instrument, not designed to ensure fulfilment of obligations through control and sanction, but by a volunteer peer review mechanism. The obligations of the Contracting Parties are mainly based on the international safety standards developed by the IAEA in past decades. The Convention is of interest of all countries generating radioactive waste. Therefore it is relevant not only for those using nuclear power, but for any country holding radioactive materials from application of nuclear energy in education, agriculture, medicine and industry, or from uranium mining and production. Obligations of Contracting Parties include attending a Review Meeting held every 3 years and prepare National Reports for review by the other Contracting Parties. In the National Reports basic information on inventory and facilities for management of radioactive materials has to be provided. Non-nuclear countries with radioactive materials under the scope of the Convention can benefit from the exchange of information and the technical knowledge gained by the reporting procedure set up by the Convention. The second Review Meeting was held at IAEA headquarters from 15 to 26 May 2006. This paper presents the objectives and the implementation status of the Convention, the expected outcome for the worldwide safety, and the benefits for a country to be part of it.

Keywords: spent fuel management, radioactive waste management, joint convention

1. Introduction

The safe management of spent fuel and radioactive waste generated by nuclear power and fuel cycle plant operation is a key issue for the use of nuclear energy. Radioactive waste is also generated whenever nuclear technology is applied in medicine, industry and research, which implies that a need of ensuring safety in dealing with radioactive material is of importance for all the countries involved in such activities, even if they do not have nor plan nuclear industrial programs. Recognizing this, the international community promoted a Convention directed to ensure that sound practices are planned and implemented worldwide for the safety of both spent fuel and radioactive waste management.

In March 1995 the IAEA Board of Governors first endorsed a proposal to convene a Group of Experts to draft the convention. The Group, made up of 128 representatives from 53 countries and observers from 4 international organizations, met seven times from July 1995 to March 1997 and drafted the Joint Convention on the Safety of Spent Fuel Management and on the Safety of Radioactive Waste Management (in short, the Joint Convention).

The Joint Convention was adopted by a Diplomatic Conference purposely convened in Vienna from 1 to 5 September 1997 and opened for signature on 29 September 1997, the first day of the 41st regular session of the IAEA's General Conference.

The Convention entered into force on 18 June 2001, i.e. 90 days after the deposit with the IAEA of the 25th instrument of ratification, as provided for by Article 40. To date the Convention has been ratified by 41 countries.

The Joint Convention is made by and belongs to the Contracting Parties. The IAEA, other than being the Depositary, provides the Secretariat and promotes the Convention, with a view to having all countries holding radioactive materials to become Contracting Parties.

2. Nature and Scope of the Joint Convention

The Joint Convention is the first international binding legal instrument in the area of nuclear spent fuel and radioactive waste safety. It is a sister convention to the Nuclear Safety Convention [1], which covers the safety of nuclear power plants, adopted in Vienna on 17 June 1994. It is incentive in nature, i.e. it does not invoke penalties for non-compliance by the Contracting Parties, but is solely based on their common interest to achieve and maintain a high level of safety in nuclear spent fuel and radioactive waste management.

The overall objective of the Joint Convention is to achieve and maintain a high level of safety worldwide in spent fuel and radioactive waste management, through the enhancement of national measures and international cooperation, so that at all stages of operation and in whatever condition, individuals, society and the environment will be protected from the harmful effects of ionizing radiation. The Convention applies to the safety of management of:

- Radioactive waste and spent fuel from nuclear power plants
- Radioactive waste from fuel cycle plant operations and from research laboratories
- Radioactive waste from the use of radionuclides in medicine, agriculture, industry and education (disused sealed sources)
- Discharges to the environment from regulated nuclear facilities
- Waste from mining and processing of uranium ores.

The Convention is practically of interest of all the countries of the world. Radioactive waste within the scope of the Convention is in fact generated not only for those using nuclear power, but by any country where applications of nuclear energy in education, agriculture, medicine and industry are currently used.

The Convention does not apply to the spent fuel held at reprocessing plant for reprocessing and does not cover waste containing only NORM (Naturally Occurring Radioactive Materials) that does not originate from the nuclear fuel cycle. Both materials can be voluntarily included in the scope of the Convention by a formal declaration of a Contracting Party.

Spent fuel and radioactive waste generated within military and defence programs are also outside the scope of the Convention, unless a Contracting Party declares such materials to be included in it, or when they are permanently transferred to and managed within civilian programs.

3. Provisions and Obligations

The obligations with respect to safety in the Convention are largely based on the principles contained in the IAEA Safety Fundamentals document "The Principles of Radioactive Waste Management" published in 1995 [2], as well as in the supporting international Safety Standards further developed by the IAEA.

The Convention contains 44 Articles. A number of them treat the legislative, regulatory and organizational framework to be established in a country in order to ensure safety, which are generally based on the requirements established in the relevant Safety Series Document of the IAEA [3–7]. Special attention is given to the transboundary movements of spent fuel and radioactive waste, in order to ensure that shipments involving two or more States take place in a manner consistent with the internationally accepted safety principles, taking into consideration the reciprocal rights of the States of origin, transit and destination. The Joint Convention is also the first binding international instrument addressing the safety of disused sealed sources, the improper use of which has raised concern among the international community. In particular, provisions are made to facilitate the return of spent sealed sources to a competent organization for reuse, storage or disposal.

Since the Convention is solely intended to stimulate improvements in safety, the fulfilment of the obligations is not based on control mechanisms but on a procedure of mutual peer review, carried out through meetings of the Contracting Parties.

A significant obligation for a contracting Party is to go through this peer review process. It consists of:

- A Review Meeting, held every 3 years by the Contracting Parties.
- A National Report, to be submitted by the Contracting Parties for review at Review Meetings.

In the Natinal Report the Contracting Party is required to explain its overall approach to the safety of spent fuel and the safety of waste management, including the existing legislative and regulatory structure, to describe policy and practices on the matter, to provide information on spent fuel and waste management facilities in operation and under decommissioning.

Provisions are established in the Convention to protect from disclosure information that a Contracting Party identifies as confidential.

National Reports are submitted by Contracting Parties prior to the Review Meetings, and distributed to all the other Contracting Parties, in order to enable their review. The Contracting Parties may seek clarification on the circulated National Reports through a written question and answer process. This peer review process is finally completed at the Review Meeting, where the Contracting Parties have the opportunity to present and discuss their National Reports.

4. Expected Impact of the Convention

There are nowadays two aspects that characterize worldwide the management of radioactive waste: the variety of safety policies and national provisions, also among countries with the same level of nuclear development; a "grey area" of activities and practices involving two or more States – like transboundary shipments, discharges, emergency preparedness or sealed sources use – for which an enhancement of the international cooperation is desirable in order to ensure the safety at a larger scale.

In addition to the above, there are a number of less developed non-nuclear countries where radioactive waste is still generated from the applications of nuclear technology in medicine, agriculture, industry and education. In these countries the lack of adequate infrastructures for the radioactive waste management, both on technical and institutional side, may lead to a non-satisfactory level of safety and radiological protection.

The Joint Convention is intended to be an instrument to effectively address the above points. Insofar as an increasing number of countries fulfil the requirements set up in the Convention, significant outcomes can be progressively achieved in radioactive materials management, in particular:

- Improved harmonization worldwide of safety policies and provisions
- Strengthening of mutual rights and responsibilities among the involved States in dealing with activity carried out internationally
- Homogeneity of infrastructures and practices worldwide for ensuring safety.

As it is well known, radioactive waste and its disposal are commonly perceived as one of the most delicate environmental problems of our time. Evidence of this concern is by the way seen in the difficulty encountered in selecting suitable sites for the final repository for this waste.

The Joint Convention, as part of the growing international effort for enhancing on a global scale the safe management of radioactive waste, is a constructive step to address the problem of the public confidence on radioactive waste practices and policies.

5. Rights and Benefits for Contracting Parties

Upon becoming Contracting Party, a country is not only subject to obligations. Thanks to the transparency of the review process established by the incentive nature of the Convention, it also acquires rights and gains benefits from it.

First of all, a Contracting Party has the right at all times to be informed about programs, policies and practices on spent fuel and radioactive waste management of any other Contracting Party whose related activity can have an impact on safety in its territory.

Benefits from joining the Convention are gained by all countries generating radioactive waste, whatever the size and the nature of their involvement in nuclear energy applications.

In particular, becoming a Contracting Party to the Convention:

- Countries with significant nuclear power programs will have benefits mainly on the political or social side. Internally, their voluntary compliance with international obligations on the safety of the management of spent fuel and radioactive waste, confirmed by a built-in international peer review, can improve the public confidence on those activities and positively affect the social acceptance of nuclear energy. Internationally, by voluntarily explaining how they meet the requirements of the Convention through the reporting process, they demonstrate at the same time the transparency of their activities on the waste management and the reliability of their technology.
- Countries with small nuclear power and/or research programs or countries having radioactive materials only from nuclear application on medicine, agriculture industry and education, can benefit from the exchange of information and the technical knowledge gained by the reporting procedure set up by the Convention, through which the expertise of larger countries is made available. Technical assistance may then be facilitated between Contracting Parties in meeting the obligations under the Convention, in particular when less developed countries are involved.

6. Status of Implementation of the Convention

6.1. THE CONVENTION'S MEETINGS

Pursuant to Article 29 of the Convention, a Preparatory Meeting was held in December 2001, attended by 27 Contracting Parties who had ratified at that date.

The Preparatory Meeting had been established to provide rules, procedures and time schedule for implementing the review process. In particular, guidelines have been agreed upon for the structure and content of the National Reports to be submitted by the Contracting Parties, and on how to conduct the Review Meetings.

Among the rules and procedures decided by consensus at the Preparatory Meeting, it is worthwhile to mention the following:

- In order to make the review of the National Reports more efficient, it has been decided to establish Country Groups for each Review Meeting, in which the National Report of each member of the Group can be considered in detail. The Groups are not made-up on a geographical base, but on a balance of nuclear power plants operated or under decommissioning in the included countries.
- An Organizational Meeting will be held seven months before each Review Meeting in which, inter alia, decisions will be taken on the mechanism to establish the Country Groups and their modus operandi, and to elect officers for the Groups and the Review meeting.

6.2. THE TWO REVIEW MEETINGS

The first Review Meeting of the Contracting Parties was held from 3 to 14 November 2003 at the Headquarters of the International Atomic Energy Agency (IAEA). It was attended by 33 Contracting Parties. The Contracting Parties were assigned to one of five Country Groups, in which oral presentation of the National Reports and discussion on written questions and answers took place.

The participation to the written questions and answers process was successful. More than three thousand questions were made totally to the 33 Contracting Parties by the others, showing the substantial interest of all countries to seek and share information on the safety of spent fuel and waste management. Technical issues and policy matter were more or less equally queried.

The second Review Meeting was held at the headquarters of the IAEA on 15–24 May 2006 [8]. 41 countries attended. China, having internally approved the Convention, but having not yet sent the instrument of ratification to the IAEA, was admitted as a full participant, by unanimous decision of the Plenary Session the first day of the meeting. Nevertheless, China did not submit the National Report. Contracting Parties and Country Groups at the second RM are indicated below:

- Group 1: Belgium, Belarus, Brazil, Croatia, Italy, the Netherlands, Romania, Spain, USA
- Group 2: Austria, Denmark, Estonia, France, Lithuania, Slovakia, Slovenia, Sweden
- Group 3: Argentina, Australia, Bulgaria, EURATOM, Iceland, Japan, Latvia, Ukraine
- Group 4: Czech Republic, Greece, Hungary, Korea, Luxembourg, Poland, Russian Federation, United Kingdom
- Group 5: Canada, Finland, Germany, Ireland, Morocco, Norway, Switzerland, Uruguay.

7. Effectiveness and Findings of the Review Process

In the second Review Meeting progress has been made since the First Review Meeting. Areas for which the need for further work was identified at the first Review Meeting have been addressed by the Contracting Parties and reflected in their National Reports and oral presentations during the Second Review Meeting.

The Contracting Parties demonstrated their commitment to improving policies and practices particularly in the areas of:

- National strategies for spent fuel and radioactive waste management
- Engagement with stakeholders and the public
- Control of disused sealed sources.

Challenges continue in a number of areas including the implementation of national policies for the long-term management of spent fuel, disposal of high level wastes, management of historic wastes, recovery of orphan sources, knowledge management and human resources.

Among the Contracting Parties there was a wide spectrum of size and scope of nuclear programs. There were Contracting Parties with major nuclear power programs, others with only hospital waste and disused sealed sources. The National Reports therefore varied appreciably in size, scope and complexity. Almost all the Contracting Parties (27 out of 41) voluntarily placed their National Reports on their public website of the Convention (Rasanet.iaea.org/conventions/waste-jointconvention.htm).

The meeting has also recognized that the fulfilment of the Joint Convention's objectives require the participation of all the countries which have spent fuel and/or radioactive waste, and that a major effort has to be made to have more Member States become Contracting Parties to the Convention. This concept was strongly emphasized in the Report of the Chairman of the Review Meeting, who also recommended the Contracting Parties and the IAEA to make efforts to have in the Convention most of the IAEA Member States by the next Review Meeting, to be held in May 2009.

8. Conclusions

The Joint Convention has been recognized by the international community as an instrument to ensure the safe management of radioactive materials worldwide. With the fulfilment of this objective, a uniform and higher degree of protection of individuals, society and environment from ionizing radiation can be achieved on a global scale. The Convention is of interest of all countries generating radioactive waste. It is in fact relevant not only for those using nuclear power, but for any country where application of nuclear energy in education, agriculture, medicine and industry is currently used. The Joint Convention is incentive in nature, which means that it is designed to be an instrument to stimulate an open self-assessment of the level safety by the countries who become Contracting Parties, through a transparent reporting and peer review mechanism allowing information and a better interaction among States on matters of safety.

The success of the Joint Convention needs a strong involvement of all potentially interested countries, both in terms of number and of "spirit" of the participation. To have more Member States become Contracting Parties is also essential for the purpose of the Convention. The IAEA Secretariat is strongly committed to this objective.

References

1. International Atomic Energy Agency (1994) Convention on Nuclear Safety, INFCIRC/449, IAEA, Vienna.
2. International Atomic Energy Agency (1995) The Principles of Radioactive Waste Management, *Safety Series* No. 111-F, IAEA, Vienna.
3. International Atomic Energy Agency (1995) Establishing a National System for Radioactive Waste Management, *Safety Series* No. 111-S-1, IAEA, Vienna.
4. International Atomic Energy Agency (1995) Safety assessment for Spent Fuel Storage Facilities, *Safety Series* No. 118, IAEA, Vienna.
5. Food and Agriculture Organization of the United Nations, International Atomic Energy Agency, International Labour Organization, OECD Nuclear Energy Agency, PAN American Health Organization, World Health Organization (1996) International Basic Standards for Protection against Ionizing Radiation and for the Safety of Radiation Sources, *Safety Series* No. 115, IAEA, Vienna.
6. Food and Agriculture Organization of the United Nations, International Atomic Energy Agency, International Labour Organization, OECD Nuclear Energy Agency, PAN American Health Organization, World Health Organization (1996) Radiation Protection and the Safety of Radiation Sources, *Safety Series* No. 20, IAEA, Vienna.
7. International Atomic Energy Agency (1996) The Principles of Radiation Protection and the Safety of Radiation Sources, *Safety Series* No. 120, IAEA, Vienna.
8. The Joint Convention on the Safety of Spent Fuel Management and on the Safety of Radioactive Waste Management, Second Review Meeting of the Contracting Parties, 15–24 May 2006, IAEA, Vienna.

CHAPTER 11

REDUCTION OF RISKS FROM LIRA UNDERGROUND NUCLEAR FACILITIES AT KARACHAGANAK OIL-AND-GAS COMPLEX

T.I. AGEYEVA, A.ZH. TULEUSHEV,
V.V. PODENEZHKO
Institute of Nuclear Physics, Almaty, Kazakhstan Republic

Abstract: The theme of this article is the investigations of radioactive contamination within and around underground cavities created by underground nuclear detonations performed in connection with the operation of oil and gas condensate fields. Underground storages of gas condensate are not maintained for a long time. The results from the large-scale complex indicate the absence of a real threat from nuclear objects on the environment. However, there is a potential danger connected with possible changes in the geological environment containing the underground storages of condensates. The pressure created in the cavities is the controlling parameter of the conditions in the cavities. Laboratory investigations of a condensate from the underground cavities confirm the absence of caesium-137 and strontium-90 and the presence of tritium. The strategy of closing off the cavities with the application of filling the cavity space with loose or helium materials is designed. The basic objectives of the subsequent works is to decrease the environmental risks associated with oil-and-gas operations and underground storage of condensates.

Keywords: cavities, underground nuclear detonations, condensate storage, LIRA, Kazakhstan

1. Introduction

Despite the fact that nuclear explosion technology used for creation of underground cavities showed high effectiveness in producing such cavities and underground containment of radioactive decay products, the concerns about potential radioactive contamination of the environment are still on the agenda [1].

The Karachaganak Oil-and-Gas Field (KOGF), Western Kazakhstan, is one of the largest oil-and-gas condensate fields in the world. An integral part of the technological flow at the field was a set of buffer underground condensate vaults (UCV) to maintain in-stock raw material availability required for the technological cycle of extraction and transportation of hydrocarbons. In 1983 and 1984, underground condensate vaults were constructed at the north of the KOGF by means of six 15 kt nuclear explosions in salt rocks, 690–840 m below the surface. These explosions are classified as "camouflet" explosions (peaceful nuclear explosions, PNEs) and the cavities are known as the LIRA facilities. Figure 1 describes the location of LIRA facilities at the territory of KOGF.

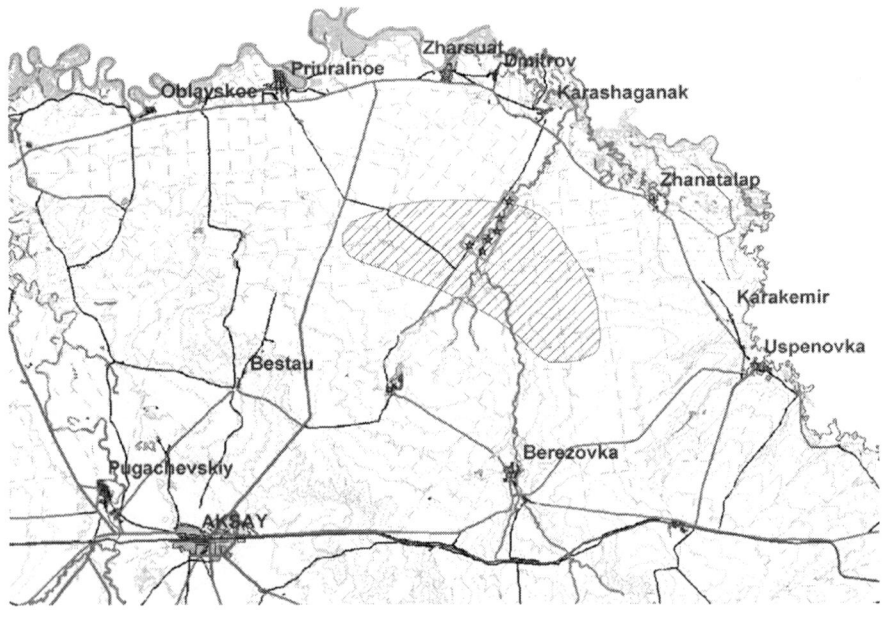

Figure 1. Location of LIRA site at KOGF

The cavities were opened and inspected after the explosions, and only four of them were found acceptable for operation; these were TK-1, TK-2, TK-3 and TK-4. Inspection of a newly bored TK-5 "bis" well showed that the cavity TK-5 was filled with water. Thus, the well was isolated by cement. In order to verify the impermeability of the water-filled cavity TK-5, a network of 11 control-and-observational wells was constructed covering the entire water-bearing strata including the salt-rock bulk, depositions of caprock and top over-salt strata.

The cavity TK-6 met the criteria for operation. However, its utilization is complex due to numerous disruptions of the bore pipes and well casing. The present work is aimed at revealing the scope of LIRA underground nuclear facilities effects on KOGF. With this end in view, the main contaminating radionuclides at the LIRA facilities were determined, as well as the amount of accumulated nuclides, taking into account their initial concentrations, by the Institute of Nuclear Physics of National Nuclear Centre in 1998 (Table I).

TABLE I. Main artificial radionuclides in underground cavities at the LIRA facilities, thier nuclear and dose characteristics

Radio-nuclides	Half-life years	Amount in 1998 Q, Bq	Dose transformation coefficient C_{tr}, Sv/Bq	$Q^* C_{tr}$ Sv
Sr-90	29	5.3×10^{13}	2.8×10^{-8}	1.27×10^{6}
Y-90	0.01	5.3×10^{13}	2.7×10^{-9}	1.23×10^{5}
Ru-106	1	8.2×10^{9}	7.0×10^{-9}	57
Sb-125	2.7	1.74×10^{11}	1.1×10^{-9}	191
Cs-137	30	5.74×10^{13}	1.3×10^{-8}	7.6×10^{5}
Sm-151	87	1.75×10^{12}	4.8×10^{-9}	870
Total		3.99×10^{14}		2.16×10^{6}

As one can see from Table I, the total activity of the main artificial radionuclides in the cavities at the LIRA site is estimated to be 3.99.10^14 Bq, corresponding to 2.16.10^6 Sv (dose transformation coefficients taken into account).

The underground cavities TK-1, TK-2, TK-3, TK-4 and TK-6 were accepted for operation, and included in the pipeline network of the Karachaganak field, as showed in Fig. 2. Except for the TK-6 cavity, the vaults were in operation for 5 years (1989–1993). Since 1994, these were technologically equipped and partially filled with condensate cavities TK-1 – TK-4 are exempted from operation, but are still kept under pressure 20–50 atm. Only in the cavity TK-6 the pressure has not been maintained for the last 20 years.

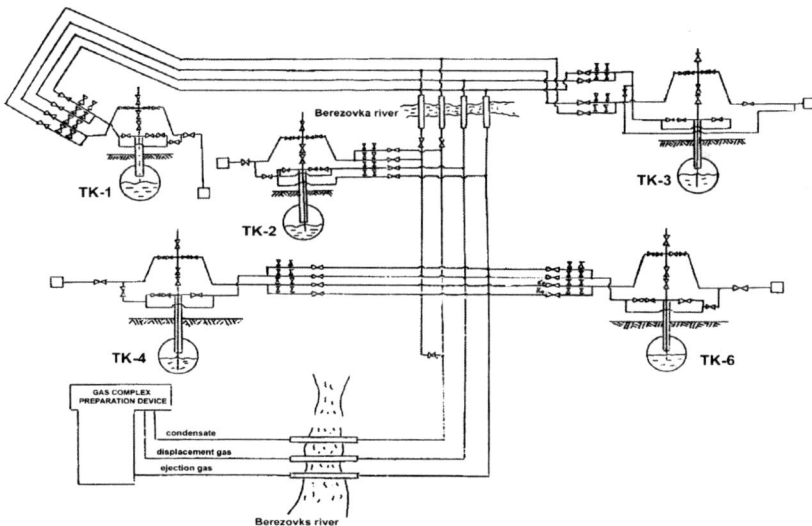

Figure 2. Infrastructure of the LIRA facilties

2. Methodology

To assure the long-term radiation safety at KOGF related to LIRA facilities and prevent their influence on oil-and-gas operations, the Institute of Nuclear Physics NNC RK has developed a concept of radiation safety of LIRA facilities and adjacent territories. Tis concept is implemented in the project "Complex investigation and monitoring of LIRA facilities". The project has been ongoing since 1998.

The main objectives of the investigations were the following:

- Radiation safety assurance, including establishing a system for complex monitoring and technical maintenance of the engineering equipment at the site.
- Reduction of risks that might influence the operations at KOGF, namely, development of technological methods to reduce the risks related to radiation and gases.

Initial assumptions at the beginning of the project were:

- Radioactive contamination has migrated over large territories and into considerable volumes of the earth crust.
- The cavities are not impermeable and thus may act as sources of radioactive contamination.

In order to obtain reliable information on the radioecological situation and evaluate future actions to be taken, a comprehensive research program for LIRA facilities and adjacent territories was developed. In addition, all required measures to assure proper technical performance of the engineering networks of LIRA facilities were undertaken, along with development of new infrastructure for monitoring and required technological operations.

In 1998–2003 a wide range of research activities were performed aiming to:

- Determine the levels and character of radioactive contamination of the near-surface and in-depth strata at LIRA sites and adjacent territories
- Determine potential pathways and migration rates for radioactive contamination from TK-5 cavity
- Evaluate the present state of the TK-6 cavity
- Develop and implement a monitoring system designed to provide reliable information on the radioecological situation at LIRA sites
- Predict the fate of the radioactive contamination based on monitoring data.

3. Main Results and Discussion

The present studies provided reliable information that significantly changed our understanding of the risks associated with LIRA facilities [2, 3]. Based on the obtained data, the current level of contamination of anthropogenic radionuclides in the soils at the LIRA sites and adjacent territories lies within the ranges which are characteristic for the

TABLE II. Concentration of Cs-137, Sr-90, and Pu-239+240 radionuclides in surface soil layers at the sites under control

Monitoring objects	^{137}Cs, Bq/kg Min–max	^{90}Sr, Bq/kg min–max	$^{239+240}$Pu, Bq/kg Min–max
LIRA site	<1.0–22.0	<1.1–4.9	0.25–0.49
Inhabited locations	0.8–15.0	<1.2–3.6	0.20–0.75
Oilfield outside LIRA site	<0.8–7.5		
Background sites	4.6–6.0	1.9–2.2	0.25–0.39

global fallout, with average specific activities of 6.6 ± 1.4 Bq/kg for ^{137}Cs, 0.30 ± 0.06 Bq/kg for ^{239,240}Pu and <5 Bq/kg for ^{90}Sr [3, 4].

As one can see from Table II, the concentrations of anthropogenic radionuclides in soils from investigated sites coincide with those at the background sites and did not exceed the level of global fallout.

The radioactive migration from the underground cavities in scales may in the foreseen future result in considerable changes of radiological situation at KOGF [5–7]. Table III presents data on the concentrations of radionuclides in water for all underground water-bearing horizons available within the site.

TABLE III. Concentration of artificial radionuclides in underground waters at LIRA site

Depth of water-bearing horizon, m	^{90}Sr, mBq/l	$^{239+240}$Pu, mBq/l	^{137}Cs, Bq/l	^{3}H, Bq/l
265	5.5 ± 1.9	<0.059	0.003 ± 0.001	<4
150	<1.9	<0.046	<0.002	<12
75	<1.6	<0.047	<0.001	<8
35	<2.0	<0.041	0.003 ± 0.001	<10
Distilled water	<2.2	<0.051	< 0.002	
Intervention level according to RSN-99	<4,500	≤500	≤9.6	≤7,700

To a large extent, the stability of the situation depends on the state of the technological equipment. To evaluate the potential risks for the personnel and the environment in cases of possible emergencies related to failure of engineering constructions, the consequences for each of the potential events were evaluated. For example, in the case of disruption of near-mouth studs at the technological well TK-4, there will be (1) a release of gas from the cavity with a jet of gas rising 26 m above the surface lasting for 1 day, (2) pollution of air with hydrogen sulfide at concentration rates 0.84–0.18 mg/m^3 and (3) depending on wind, the pollution will expand for 1–3 km. Consequently, proper maintenance of the technological equipment is a strict requirement, and related works are now performed in compliance with the requirements set for the oil and gas industry.

The underground condensate vaults TK-1–TK-4 have not been in operation since 1994. With the purpose to control their technical conditions the underground vaults and their technological wells are kept under pressure. Pressure in the vaults was provided by pumping crude gas from the complex gas preparation installation CGPI-3 (UKPG-3) into the vaults. Since the gas contains such reactive components as hydrogen sulfide and carbon dioxide, anti-corrosive treatments of the piping were performed annually (even after operation of the vaults terminated). However, no measures have been applied to prevent corrosion since 2002 due to the termination of the gas supply from the gas field. Studies revealed that, to different extents, inner surfaces of all parts of the technological equipment are influenced by corrosion. Changes in pipe wall thickness close to the critical value were revealed in parts of the pipe system. In addition, deterioration of locking units results in loss of permeability which is indicated by gas leaks in the near mouth equipment. Thus, the most important issue today is to lower the risks related to the gas hazards. Considerable progress in this reduction may be achieved by pressure release in the underground cavities. However, there are also certain risks related to such a release (Table IV).

TABLE IV. Risk factors and expected consequences at LIRA facilities

Risk factor	Expected consequences
Pressure of rock formation on vault roof	Collapse of vault roof
Plasticity and creep of salts	Change in geological situation, filling-in of cavities with salt
Presence of a zone with open well shaft between the bottom of technical well and upper vault	Contortion of the operational well and emergency due to collapse or compression by salt rock creep
Break up of cementation ring between casing strings and surrounding rocks	Infiltration of water from water-bearing horizons of Triassic and Neogene-Quaternary strata with filling up the cavity
Tectonic phenomena	Collapse of underground vault, filling-in with water and formation of radioactive brine

Based on the factors stated above and their possible consequences, it is obvious that disintegration of abandoned technological constructions with time in the present geological conditions is probable.

An interesting geological stability of a media in post-explosion conditions is presented by the TK-6 cavity that was left at no pressure for almost all time since its construction (except short periods of purging and overhaul). Special studies performed in 2004 showed that the well shaft was easily assessable down to 800 m with no water in it, while access is currently impossible due to technological obstacles in the well shaft. Geological and geophysical investigations of the geological structures in the vicinity of this cavity showed that the geometry of the maximum fissuring zone around the cavity is far from spherical (Fig. 3) demonstrating considerable anisotropy in physical and mechanical properties of the geological media enclosing the cavity.

The spatial location and the shape of the main fractioning zone around the cavity TK-6 were determined by means of seismic investigations, in order to reveal the morphology of presumably weakened zones related to the cavity. The presence of softening regions in the geological bulk is a potentially hazardous factor since such zones may provide migration paths for radionuclides. Currently there is no information about conditions inside the cavity.

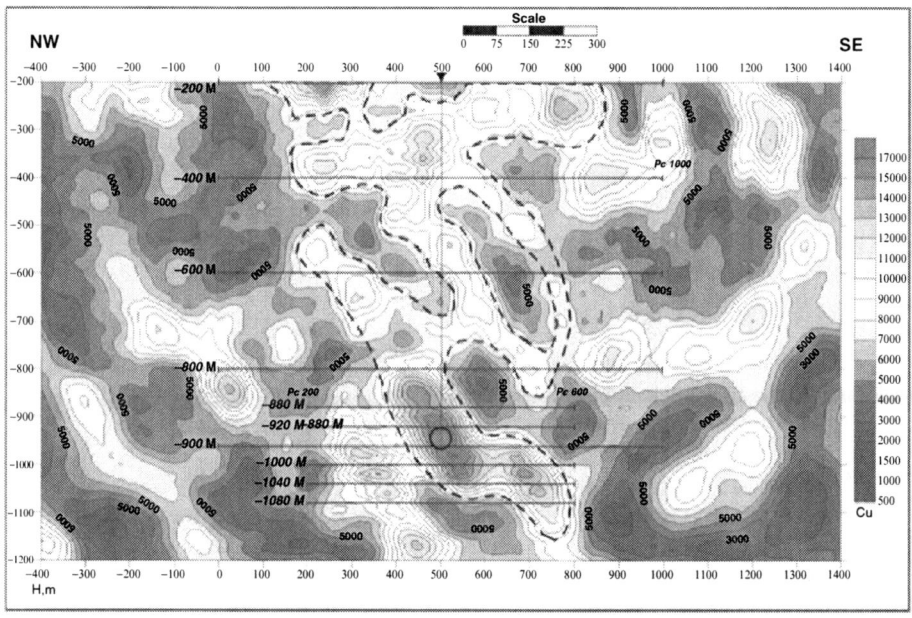

Figure 3. Results of seismic investigations of TK-6

Based on stable readings of pressure gauges at the mouths of technological wells TK-1–TK-4 during 13 years after ceasing the UCV operation, one can say that the geological formation around the cavities is currently stable and the underground cavities are sealed. But the processes in geological strata are of long-term nature and they can be triggered by any phenomenon like natural or artificial seismological activity. Thus, mouth pressure monitoring is required to control the conditions in the cavities. It is reasonable to leave the cavities at no extra pressure for some time prior to any measures on "liquidation" of the cavities.

Taking into account the radiophobia and public concerns, as well as rejection of the investors to use the UCVs in the technological process, the general strategy to close the cavities in order to prevent migration of radionuclides has been outlined. The main step in the procedure of closing the vaults is to fill up the vaults with particulate natural material, solid industrial waste or crushed ore mixed with sorbents. The Institute of Nuclear Physics NNC RK is running laboratory investigations and pilot studies to identify and characterize types of filler materials and sorbents that may be used for this purpose. Liquid and gaseous hydrocarbons in the cavities need to be removed prior to closing the cavities. However, the radionuclide concentrations in the hydrocarbons to be removed need to be within permissible concentration levels. Sampling of condensate from the vaults TK-1, TK-2, TK-3 and TK-4 was performed in August 2006 to evaluate the radionuclide contamination. Laboratory studies of the condensate samples showed radionuclide concentrations (^{137}Cs and ^{90}Sr) to be very low, and in agreement with investigations performed at similar facilities of Orenburg oblast in Russia [8]. Currently the INP runs laboratory investigations to reveal the concentration of gaseous radionuclides in the condensate. Preliminary results show the content of tritium to be 1.5–2.10^3 Bq/l in one of the vaults.

4. Conclusion

The current experience and knowledge on the levels and characteristics of the radioactive contamination at the LIRA facilities, adjacent territories and the background level at the site location, makes it possible to prevent the impact from the LIRA facilities on oil industry at KOGF; there are all the grounds to evaluate the scales of the required works at acceptable achievable level. The main tasks directed towards reduction of risks for the oil industry and the environment related to the underground nuclear cavities are as follows:

- To implement comprehensive monitoring in order to provide information on the state of the LIRA facilities and any changes related to them.
- To improve the competence and resources related to maintenance and elimination of risks due to gas hazards.
- To develop technology for hydro-isolation of water collectors and the emergency response system in case of radionuclide migration from the water-filled cavity TK-5.
- To investigate the present state of the TK-6 vault.
- To drill a well in the vicinity of TK-6 well for observation purposes.
- To further develop the technological methods and the economical basis for the closing of underground vaults.

5. Acknowledgement

The authors express their gratitude to S.N. Lukashenko, INP, for data used in this paper.

References

1. Logachev V.A. (2001) Peaceful nuclear explosions: general and radiation safety assurance at explosions. *M. IzdAT*. 519 pp.
2. Kadyrzhanov K.K., Tuleushev A.Zh., Lukashenko S.N., Solodukhin V.P., Kazachevsky I.V. and Reznikov S.V. (2001) Analysis of nature and mechanisms of contamination with radionuclides at LIRA site. *Proc. 3rd International Conf. "Nuclear and Radiation Physics"* June 4–7, 2001. Almaty, Kazakhstan.

3. Silachev I.Yu., Podenezhko V.V. and Lukashenko S.N. (2003) Origin and distribution character of 239+240Pu at LIRA site and adjacent territories. *Proc. 4th International Conference "Nuclear and Radiation Physics"* Sept. 15–17, 2003. Almaty, Kazakhstan.
4. "Report-Substantiation for changes in periodical reporting on sampling and checks at LIRA facilities and on adjacent territories"/a group of authors from the Institute of Nuclear Physics NNC RK, Almaty, INP NNC RK 2004.
5. Lukashenko S.N., Ageyeva T.I., Solodukhin V.P., Melentiev M.I., Kisliy B.I. and Zholdybayev A.K. (2001) Revealing of potential migration pathways within rock formations with the cavities formed by underground nuclear explosions according to geophysical and geochemical data on example of Balapan site and LIRA site. *Proc. of the 3rd International Conference "Nuclear and Radiation Physics"* June 4–7 2001. Almaty, Kazakhstan.
6. Ageyeva T.I., Tuleushev A.Zh., Marabaev Zh.N., Lukashenko S.N., Reznikov S.V., Novozenko V.A. and Borisenko A.N. (2002) Influence of the LIRA facilities on radiation conditions of the adjacent territories: rezalts of the four-year monitoring. ECNSA 2002 *Advanced Research Workshop Environmental Protection Against Radioactive Pollution*, Sept., 16–19, 2002, Almaty, Republic of Kazakhstan, ABSTRACTS. Almaty, Institute of Nuclear Physics NNC RK, 116 p.
7. Ageyeva T.I., Tuleushev A.Zh., Lukashenko S.N. and Reznikov S.V. (2002) Radio-ecological situation at LIRA site and Karachaganak oil-and-gas field. *KazNU Bulletin*, Chemistry series No. 6 (Special issue), Almaty, Kazakhstan.
8. Dubasov Yu.V., Trifonov V.A., Smirnova Ye.A. and Arshanskiy S.M. (2005) Present state at the territory of peaceful nuclear explosions in Orenburg oblast. *Radiochemistry*, 2005, **47**, No. 6, pp. 556–563.

CHAPTER 12

THE NET EFFECT OF THE ARMENIAN NUCLEAR POWER PLANT ON THE ENVIRONMENT AND POPULATION COMPARED TO THE BACKGROUND FROM GLOBAL RADIOACTIVE FALLOUT

K. PYUSKYULYAN[1], V. ATOYAN[1],
V. ARAKELYAN[2], A. SAGHATELYAN[3]
[1] *Armenian Nuclear Power Plant, Medzamor, Republic of Armenia*
[2] *The Yerevan Institute of Physics, Yerevan, Armenia*
[3] *Institute of Noosphere Researches of AS of Armenia*

Abstract: An evaluation of the direct nuclear radiation effect which Armenian NPP may have induced on the population living in the NPP location area, as well as on the local environment, compared to the background from global fallout from nuclear weapons test and fallout from the Chernobyl accident is presented. The concentration of ^{137}Cs in soils in the territory with a radius of 50 km, including the area near Armenian nuclear power plant, has been investigated. Meteorological and hydrogeological characteristics of this territory are studied in detail. The data describing radiation conditions in the region where the Armenian nuclear power plant is situated, during the investigation period and for the period from 1978 to 2004 are cited.

The evaluation is made by two independent methods. On the basis of the analysis of meteorological data and calculations of the distance to which radioactive emissions from the nuclear station have the maximum concentration in air, the monitoring points were chosen. By means of direct measurement of the concentration of cesium in soil at these points and comparative analysis, the direct contribution from the nuclear power plant to radiation conditions in the environment was determined. The obtained data, describing the net effect of the nuclear power plant on the population and environment, are well coordinated with data of the United Nations Scientific Committee of Radiation Action.

Keywords: ^{137}Cs, soils, microorganisms, nuclear power plant, Armenia

1. Introduction

For an estimation of the radiation risk connected with operation of the nuclear power plant (or other objects allocating technogenic radionuclides in an environment) it is necessary to define a radiation impact level of such objects on an environment. The aim of this research was to identify the net radiological impact of the Armenian NPP on population residing in its surrounding area and on the local environment.

2. The Approach to Measure the ^{137}C Distribution in Soil

In cases when the radionuclide emissions into the environment is small, the identification of the NPP impact is difficult as there is a necessity to compare the activity levels observed with that of the background from the global fallout (consequence of atmospheric nuclear weapons tests performed by the different countries during a certain period). The magnitude of the global fallout is much higher than fallout from the NPP. In some cases, it is also necessary to consider the fallout from the Chernobyl accident.

In our researches the ^{137}Cs was chosen as a main investigated radionuclide. Soiø was selected as the major elemental compartment as ^{137}Cs accumulates in soils. Naturally occurring ^{137}Cs is a product of spontaneous fission of ^{238}U isotope. According to different authors, the specific activity of natural ^{137}Cs in soil under stable conditions varies from 3.7 to 370 μ Bq/kg. This value is 4–5 orders less than the concentration of all ^{137}Cs in soil. Hence, the concentration of cesium in soil at the NPP location first of all is caused by global fallouts, and also (in much smaller degree) by the influence of the nuclear power plants. ^{137}Cs is a major dose-forming radionuclide in the global fallout. Its contribution to the total dose to the public is nearly 40% [1, 2]. Transfer of radioactive isotopes supplied to the troposphere and stratosphere in course of nuclear tests, as well as their fallout on the Earth's surface are sufficiently well studied [2].

The following approach was used to determine the distribution of ^{137}Cs isotope in air and in soils. The concentration of aerosols emitted by the NPP ventilation pipe is known to be equal to zero at a point closest to the source on the ground. With distance the concentration increases, reaching maximum at a distance (X_m) depending on the weather conditions and state of the gas-aerosol mixture, and then the concentration decreases. Dispersion of an impurity in an atmosphere is precisely enough described by means of the Pasqill Setton Main Geophysical Observatory techniques. Calculations performed by means of the techniques have been published [3]. We have calculated the distance (X_m) from the ventilation pipe to the point of maximum concentration, with account of the weather conditions in the location area of NPP and the technological characteristics of the emitted gas-aerosol mixture (the main monitoring point M-1). The Armenian NPP is situated in the Ararat valley where some primary wind directions (see Fig. 1) are observed. The maximum concentration of the ^{137}Cs isotope is expected in fallout at the point M-1, as well as in the soil. Thus, maximal impact of the fallout from the NPP should be expected here (uniformly distributed in the observation area).

Soil samples were collected during the period May 2003 till May 2005. A special frame made of steel was used, 10 by 15 cm and 5 cm in height. Each sample included five frames selected by an envelope method. Up to 35 samples were collected in each point during the experimental period. The selected soils were vegetation-free and similar in type, mostly brown semi-desert soils. Besides, as background tests, we collected soil samples from the points remote from the Armenian NPP, at a distance of 60 km from the site. The soil type in this area corresponded to the type of soil in area surrounding the Armenian NPP. The collected soil samples, according to the generally accepted methods, were initially dried, crushed and then weighed. Gamma-spectrometric analysis of samples was performed using a low background installation with pure germanium detector (Canberra production) and program support "GENIE-2000".

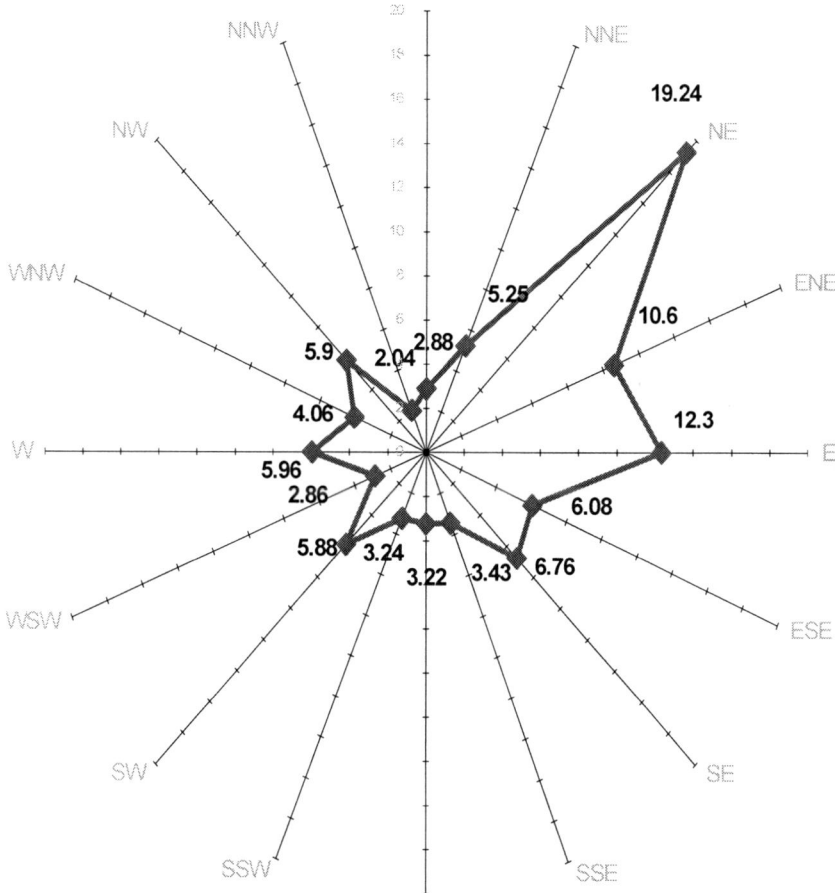

Figure 1. Wind directions in the ANPP industrial area, averaged over the period 1970–1998

The gamma-spectrometric analysis of the samples showed that the soil contained natural radionuclides such as ^{40}K, ^{226}Ra and ^{232}Th, and also ^{137}Cs (the relative error of measurements changed within the limits from 1.5% up to 3%). In Table I and Fig. 2 the concentrations of the radionuclides are given.

TABLE I. Annual mean concentrations of radionuclides in soil at the monitoring points

Monitoring points	Distance from NPP	^{137}Cs Bq/m^2	^{40}K Bq/kg	^{226}Ra Bq/kg	^{232}Th Bq/kg
Average results for the first year					
Agavnatun	10 km on NE	476	534	25.2	35.8
Point M-1	2.2 km on SW	586	434.3	24.8	29.5
Metsamor	5.0 km on SW	565	447	23.2	47
Armavir	10 km on SW	488	445	30.4	25.8
Nor-Armavir	17 km on SW	454	308.7	29.3	63.1
Average results for the second year					
Agavnatun	10 km on NE	571	514	26.7	38.4
Point M-1	2.2 km on SW	674	448	21.5	31.2
Metsamor	5.0 km on SW	621	483	25.3	42.6
Armavir	10 km on SW	559	461	28.6	27.5
Nor-Armavir	17 km on SW	550	389	14.8	51.4
Average results for the third year					
Agavnatun	10 km on NE	489	497	29.2	42.5
Point M-1	2.2 km on SW	617	465	24.7	34.2
Metsamor	5.0 km on SW	582	491	26.6	41.2
Armavir	10 km on SW	510	472	30.1	32.1
Nor-Armavir	17 km on SW	487	420	29.6	50.2

According to the earlier stated hypothesis, the maximum concentration of ^{137}Cs in soil was expected at a point M-1, where the maximum quantity of the emission generated by the nuclear plant should be superimposed on the global fallout (uniformly distributed in the whole observation zone). Analysis of chart in Fig. 2 shows that this hypothesis holds: the cesium concentration in soil at this point, located leewards from the Armenian NPP (Aghavnatun) where there is no impact of the ANPP releases, was less then in point M-1 and Metsamor. Namely, the ^{137}Cs concentration was highest in the point M-1 and it falls off as the distance from the plant is increased. So, it is possible to confirm, that, according to the data presented in Fig. 2, the difference in the cesium concentration in soil, caused by the influence of the emissions from the NPP, decreases when the distance increases. A difference between the ^{137}Cs concentration in point M-1 and in the background points (NPP impact) was much higher than the measurement error; it equals to 115–120 Bq/m^2 and is about 20% of the background concentration value caused by global fallout.

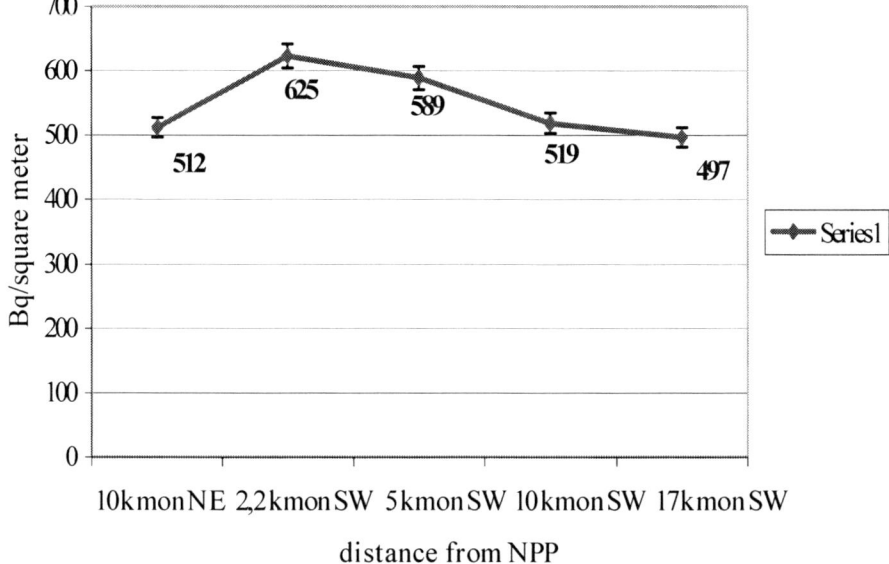

Figure 2. ^{137}Cs concentration in soil in monitoring points (average for all the period of monitoring)

3. Numerical Methods

The second way of defining the NNP impact on environment consists is to use numerical methods. The following problem was formulated: on a noisy background define a useful signal, with size smaller than fluctuations of a background.

The decision of such problem became possible as we have a large databank describing the ^{137}Cs concentration in soils for the period from 1978 to 2005 in a monitoring area with the radius of 30 km.

It was assumed that the cesium concentration in the soil (Φ_{cal}) is caused by three factors: global fallout (nuclear weapons tests), emissions from the nuclear power plant and fallout from the Chernobyl NPP:

$$\Phi_{calc} = A_{test} \exp[-\lambda(t - t_{01})] + A_{Chern} \exp[-\lambda(t - t_{02})] + \Phi * [F_{em}(t)] \quad (1)$$

Numerical calculations have been made, using the data on concentrations of ^{137}Cs in soils and its emission during the whole observation period (1978–2002). The results of these calculations are given in Fig. 3.

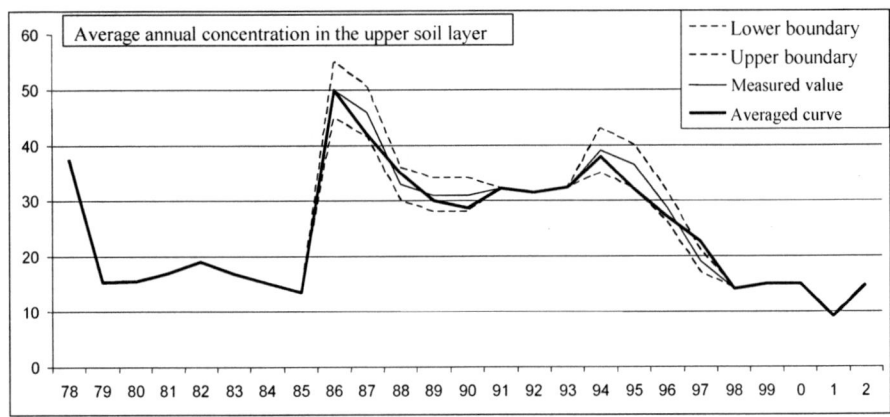

Figure 3. Calculated values of the ^{137}Cs content in soil

Here one curve is the annual average measured values of the ^{137}Cs in soil, while the another smoothed curve represents calculated values. One can see from this figure that the calculated and observed data are in good agreement. The contribution of the ANPP emissions is described by the following function:

$$\Phi_{res}[F_{em}(t)] \quad (2)$$

This function allows the evaluation of a net concentration of the ^{137}Cs in soil caused by NPP operations. Depending on the season, this input is about 14–18% of the total ^{137}Cs concentration in soils, or 90–100 Bq/m^2.

On the basis of the obtained data on the contribution of emissions from the nuclear power plant to the ^{137}Cs concentration in soils, the exposure doses of the population from NPP emissions has been calculated (Table II).

TABLE II. Exposure doses from ^{137}Cs

Monitoring points	Exposure dose from ^{137}Cs (global fallouts and ANPP emissions μSv/year)				
	External	Inhalation	Peroral	Σ	The mid-annual measured external exposure doze in monitoring area = 1314 μSv/year
2.2 to SW	1.142	0.000012	15.7	16.84	
5 to SW	1.038	0.000012	15.7	16.74	
Background	0.907	0.000012	15.7	16.61	
The dose only form NPP in max point	0.235	0.000012	3.14	3.375	

4. Conclusion

To evaluate the impact from the nuclear power plant on an environment and the population in the surrounding area, it is necessary to use special techniques and long-term monitoring of the area where the plant is located.

Data analysis of the ^{137}Cs concentration in soils has shown that, in the prevailing wind directions, the concentration of ^{137}Cs in soils is at maximum at a certain point, and varies from minimum to maximum with the distance from the NPP. The difference between the ^{137}Cs concentration at M-1 and in the background points was equal to 6–7 Bq/kg, or 115–120 Bq/m^2, which is nearly 20% of the background concentration caused by global fallout.

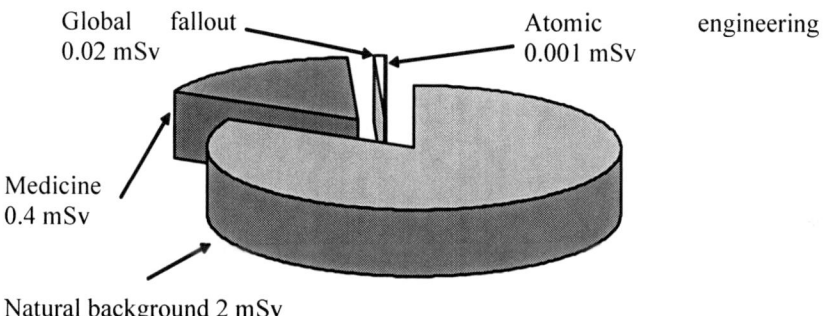

Figure 4. Average annual effective equivalent doses from natural and technogenic radiation sources according the UN Scientific Committee on the Effects of Atomic Radiation data

The total dose of an external and internal exposure of the population caused by the emissions from the Armenian NPP in a point having the maximum influence, equals 3,375 µSv/year and is about 20% of the doses from the global fallout and about 0.26% of doses from external exposures caused by the natural background. The obtained data agree well with data published by the UN Scientific Committee on the Effects of Atomic Radiation report [4] (Fig. 4).

References (in Russian)

1. Mouseev A.A. and Ivanov V.I. (1984) Dosymetry and radiation medicine reference book, Moscow, *Energoatomizdat*
2. Mouseev A.A. (1985) Caesium-137, environment, people, Moscow, *Energoatomizdat*
3. Teverovskiy E.N., Artem'eva N.E.H.E. et al. (1980) The possible radioactive and chemical air venting, Moskow, *Atomizdat*
4. UN Scientific Committee on the Effects of Atomic Radiation, Report (1988) Radiation, douses, effects, risks, Moskow

CHAPTER 13

ASSESSMENT OF RISKS AND POSSIBLE ECOLOGICAL AND ECONOMIC DAMAGES FROM LARGE-SCALE NATURAL AND MAN-INDUCED CATASTROPHES IN ECOLOGY-HAZARD REGIONS OF CENTRAL ASIA AND THE CAUCASUS

A.N. VALYAEV[1], S.V. KAZAKOV[1], A.A. SHAMAEVA[1], O.V. STEPANETS[2], H.D. PASSELL[3], V.P. SOLODUKHIN[4], V.A. PETROV[5], G.M. ALEKSANYAN[6], D.I. AITMATOVA[7], R.F. MAMEDOV[8], M.S. CHKHARTISHVILI[9]
[1]*Nuclear Safety Institute of the Russian Academy of Sciences (IBRAE RAS), 52 b. Tulskaya str., 115191, Moscow, Russia*
[2]*Vernadsky Institute of Geochemistry and Analytical Chemistry of RAS, 19 Kosygin str., 11999, Moscow, Russia*
[3]*Geosciences and Environment Center Sandia National Laboratories, Cooperative, Monitoring Center, USA*
[4]*Nuclear Physics Institute of National Academy of Sciences, 490000, Almaty, Kazakhstan Republic*
[5]*Technical University, 492000, Ust-Kamenogorsk, Kazakhstan Republic*
[6]*Yerevan State University, Department of Geology, 1 Alex Manoogian str., 0025, Yerevan, Armenia*
[7]*Institute of Physics and Mechanic of Rock Stones of National Academy Sciences, Bishkek, Kyrgyzstan Republic*
[8]*Geology Institute of Azerbaijan National Academy of Sciences, 29a H. Javid av., AZ1143, Baku, Azerbaijan*
[9]*Scientific Center for Radiobiology and Radiation Ecology, Georgian Academy of Sciences, 51 Telavi Str., 0103, Tbilisi, Georgia*

Abstract: Various threats to civilization have recently been observed to increase in number. They include natural and man-induced catastrophes, international terrorism, ecological imbalance, global climate change and many other hazards. According to UN and scientific forecasts, an increasing tendency for the catastrophe scale will retain in XXI century. The humankind has faced the majority of the above issues for the first time and, therefore, there are no good suitable methods provided for their solving. A purposeful activity of all states of the world community is required. This paper presents an international Program with the participation of six countries: Russia, Kazakhstan,

Kyrgyzstan, Georgia, Armenia and Azerbaijan. The Program includes separate Projects, developed by each participating country and is briefly described in the present paper. Today, one of the possible methods to analyse the problem is the assessment of risks and the possible ecological and economic damages resulting from catastrophes. We will pay attention to the most ecological dangerous regions and single out typical factors that significantly increase the risk of natural and man-induced catastrophes. The prediction, prevention and assessment of the above mentioned treats are especially actual and important for new independent states, where the system of safety, boundary and customs control, that of strict visa control and other state safety measures have not yet been formed. Also it is easier to implement risks associated with terrorist attacks in these regions. Their consequences will be followed by major human and huge material losses, and extremely negative irreversible environmental effects of global scale. The Nuclear Safety Institute of Russian Academy of Sciences (IBRAE RAS) has their own experience within these scientific field (http://www.ibrae.ac.ru).

Keywords: geohazards, man-made hazards, risk assessments, Central Asia, Caucasus

1. Introduction

Today, increased number of catastrophic events is notable, which demonstrates that the nature of their occurrence is far from being incidental. Not only the population is suffering, but also economic damages supported by negative or often irreversible environmental impacts are increasing abruptly and constantly. Catastrophe risks have increased so much that it becomes evident that none of the states is able to manage them independently. The situation is aggravated by the fact that a number of catastrophes may be caused by deliberate attacks of terrorists. Therefore, it is necessary to develop and implement new special methodologies, techniques and approaches aimed to forecast large natural and man-induced catastrophes as well as to prevent and eliminate their consequences. Also, it is important to notice that it is impossible to model all types of large possible catastrophes and thus to predict their consequences, as it may be done, for example, in physical experiments. In our opinion, one of the most effective ways to solve the issue can be the assessment of risks and ecological-economic damages from catastrophes. Opportunities for any country in this field are always limited, but the integration with other states will allow not only to assess but also manage transboundary risks, which is an essential prerequisite for sustainable development of any country and its transfer to a higher management level.

The latest major man–made disasters in Europe and Siberia, including sources of radionuclides and toxins in northern marine environments are presented in Fig. 1. For several years IBRAE RAS has been running system research on ecological risks, associated with environmental contamination and its health effects (http://ibrae.ac.ru). The methodology created is recommended by international organizations, such as WHO, UNEP, Russian Health Ministry and used presently in Europe and USA. Models are developed for migration of contaminants into the environment and their entrance into human body, as well as methods for estimating risk factors, including comparative analysis for radiation and chemical risks. Methods are being developed to counter some kinds of radiation terrorism, which can be provided by means of explosives [1]. We pay great attention to the regions, where it is possible to stimulate natural calamities by such actions or where man-induced catastrophes can cause huge negative effects on international or global scale. In the present work, we will present the main details of our international program. It includes separate thematic projects, developed by nine scientific organizations from the six participating countries in Europe and Central Asia: Russia, Kazakhstan, Kyrgyzstan, Armenia, Azerbaijan and Georgia. We combine our efforts and utilize the experiences of the individual institutes in solving the above mentioned problems. Brief characteristics of these regions and some of the results from the risk assessments are given below and in thematic publications [2–9].

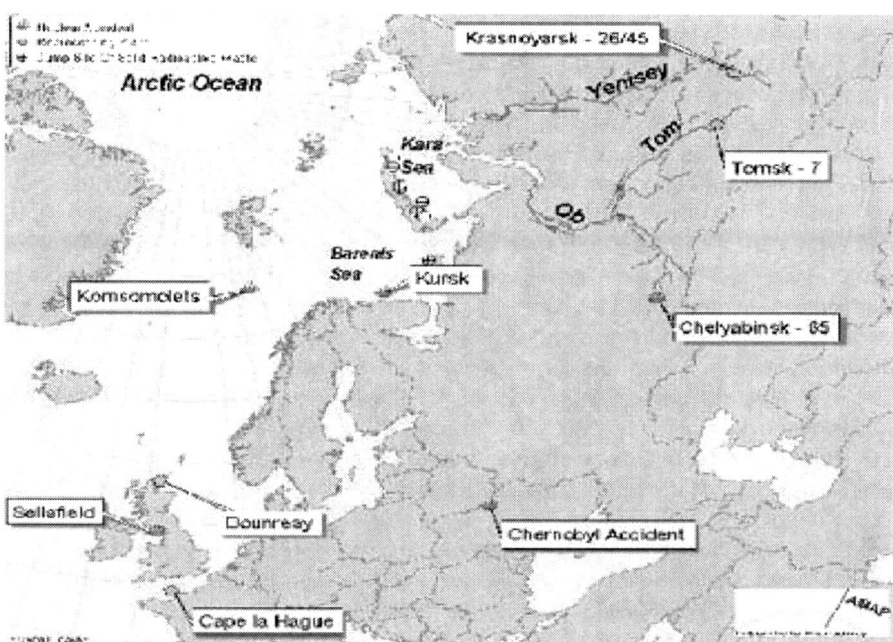

Figure 1. Sources of radionuclides in northern marine environments: Chernobyl in Ukraine, the reprocessing plants at Sellafield, La Hague and Dounreay, the dumping sites of nuclear waste in Kara Sea, Russian nuclear installations (Mayak [Chelyabinsk-65], Tomsk-7 and Krasjonoyarsk-26) releasing radionuclides to Russian Ob and Yenisey rivers and the sunken submarines Komsomolets and Kursk

2. Central Asia

Practically, the entire 4 million square kilometres territory of Central Asia's states, populated by nearly 50 million people, is an area of ecological risk of global scale. We consider the two most hazardous regions:

1. The territory in the basin of the Naryn-River, Kyrgyzstan.
2. The industrial areas of Eastern Kazakhstan, located along Irtysh River, the largest river in the Republic.

2.1. KYRGYZSTAN REPUBLIC

The largest uranium, antimony and mercury deposits in the former USSR are located at Kyrgyzstan territory in ~1,200 ha. The 49 tailing dumps, multiple burrows and slime basins comprise of about 100 million cubic metres of fine disperse radionuclide waste; heavy metal salts and toxins. High radioactive waste of enriched U-productions from Eastern Europe and China were also delivered here for the last decade. Central Asia's main upper rivers Syrdarya and Amurdarya have been contaminated and continue to be contaminated, along with transboundary contaminant transfer to areas in Kazakhstan, Uzbekistan and Tajikistan, entailing the growth of economic destabilization and political tension. The status of the majority of tailing dumps is extremely poor, as their dams have started to break with time under the impact of landslides, mudflows and ablation. Here, we pay attention to problems associated with the most dangerous KR geoecological region with U-tailing storages, located along the largest rock Naryn River, flowing into Syrdarya river (Fig. 2). Naryn River provides the whole country with energy supply, where the six major hydroelectric stations (HES) and their huge water reservoirs are located. The Torkogul HES reservoir has accumulated 20 billion cubic metres water with a ~200 m high dam. These huge water masses in HES reservoirs stimulate the frequency of more intense earthquakes, for example, as happened recently in Sysamyr [1☆] (1992, by the Richter scale M = 7.3) and in Kochkor–Ata [2☆] (1992, M = 6.2), The locations of the earthquakes are shown in Fig. 2. Detailed analysis showed that high concentrations of radionuclides such as radon, radium and different heavy metals were present in water and in materials of walls in the near located buildings and artificial constructions. In the river water pollutant concentrations exceeded 1–2 orders the extreme limited concentrations. The most difficult situation is in the territory along the Mailuu Su River and its tributaries. where 13 U-tailing sites are located. Previously, radioactive wastes from the enrichment U-productions in Earth Germany, Bulgaria, Czech and China were delivered here for second processing. Today, the following element concentrations have been registered: 0.1–0.15% U; 0.11–0.18% Cu; 0.5–1.0% Mn; 0.026–0.036 G/kg Se; 0.02–0.034 G/kg Co. The total alpha activity of the wastes varied in the range 67,000–113,000 Bq/kg. The average gamma intensity at the tailing surfaces was 20–50 µR/h, but at some places the value reached 250–2,000 µR/h. The average radon concentration reached 400 Bq/m^3 at walls in old buildings.

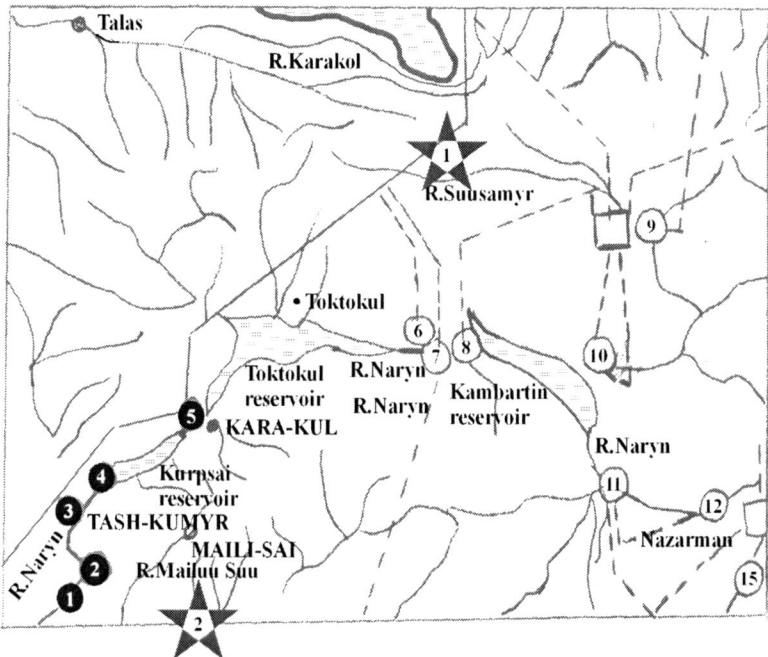

Figure 2. Objects in the Tien-Shan mountains, included in the HES cascade of the Naryn River, the largest river in Kyrgystan Republic • – Operative HES; ○ – projected HES; — – operative high voltage lines;- projected ones [1★] Sysamyr , [2★] Kochkor–Ata

The huge water masses provide a pressure on the Earth' crust surface and stimulate the frequency and the intensity of earthquakes. Besides the dams of reservoirs and tailing dumps, Russian and American military air bases, located in Kyrgyzstan are very attractive for controlled terrorist acts with usage of explosive [1].

2.2. EASTERN KAZAKHSTAN

This is also an area of major ecological risk, where multiple mines for extracting metals and minerals, as well as a number of industrial enterprises and their tailing dumps, including uranium, are located in the cities and their suburbs along the Irtysh River, the largest river of the country. Irtysh River originates in China, flows via the lands of Kazakhstan and Russia, including such large cities as East-Kazakhstan capital Ust-Kamenogorsk, Semipalatinsk, Pavlodar, Omsk, and Tobolsk After its confluence with the Ob River, it flows into the Arctic Ocean (Fig. 3). The largest Bukhtarma HES with the 100 m high dam has the sizeable reservoir; the basin length is above 300 km and the depth is up to 100 m. The other Ust-Kamenogorsk HES with a 42 m high dam and a sole hatch is located 15 km from the city.

The next Shulba HES (1) is located 180 km from Ust-Kamenogorsk. Errors in the design of these HESs have caused a number of different accidents. Unfortunately, the data on observations and statistics associated with the HESs accidents are practically absent. The Ust-Kamenogorsk city with 330,000 inhabitants is one of the most contaminated towns in the world and is located in Irtysh mountain valley. It represents a unique urban system, "oversaturated" by different enterprises. The Ulba Metallurgical Plant (UMP), the largest in former USSR, incorporates three separate works, producing enriched U for nuclear power plants, Be, Ta and their products. The operating UMP wastes storage, located in city centre, has accumulated to ~100,000 t of wastes, containing U, Th and their decay products. Its size is about 400×220 m^2 with the depth of the contamination of more than 5 m. The level of gamma radiation at its surface reaches 360 µR/h and increases with depth up to 1,000 µR/h. Radioactive anomaly regions with 1,000 and up to 6,000 µR/h are registered at UMP territory. Many other operating large city plants, such as Lead–Zinc, Titan-Magnezium, Ceramic plants, worked on Be base, power capacitors plant, nonmetalliferous group of enterprises and silk cloth enterprises, use in their technologies different poisonous chemicals and toxins, while their wastes are also located close to city. For instance, the Lead–Zinc plant stores in open cast dumps of about 17.5 ha more than 13 million tons of wastes and ~1,000 t of arsenic in the form of highly toxic substances calcium arsenate and arsenite, where 7–10% As is contained. In Irtysh River basin, where >40% of the HES energy in Kazakhstan is produced, large active non-ferrous pits, precious and rare-earths metals pits with their dump sites are also located. In Ulba River water, flowing into Irtysh in the city, the contaminate concentration is: Cu (4.86–5.50) times the maximum permitted concentration (MPC), Zn 4.71–5.37MPC, oil products 2.03–2.07 MPC, and nitrite nitrogen 1.40–1.95 MPC.

Risks of ecologic catastrophes are increased, because two HES are placed at Irtysh upriver of the city (Fig. 3). The huge water masses in the man-made seas, making a pressure on the bottom of mountain surfaces, disturb and deform their initial natural states. We consider that these factors may increase the frequency and intensity of the strong earthquakes, included catastrophic ones, that already have happened not only near the city (in 1990), but also in Altai mountains in Russia (in 2003 and 2005). Such earthquakes may cause damages on the HES dams, where in addition part of them are in non-satisfactory states, especially the Ust-Kamenogorsk HES dam, operated for more than 50 years. Also, any HES with huge water reservoirs are very attractive for planned terrorist acts, including the use of explosive. According to some primary estimates, a huge break-through damming wave with a front height of ~30 m will destroy the city and its environs. All enterprises, their products and hazard impurities in storages sites, can be carried downstream Irtysh to many cities and after Irtysh–Ob rivers junction spread over large territories, including the Arctic Ocean via the Kara Sea (Figs. 1, 3).

Irtysh basin has accumulated 120 million cubic metres of different wastes, which is 60% of the total pollution of all water basins in Kazakhstan, and can influence the water quality in all cities: Ust-Kamenogorsk, Semipalatinsk, Pavlodar, Omsk, Tobolsk and in many inhabited localities. Irtysh-Karaganda man-made channel supplies water to Kazakhstan central regions, such as Karaganda, Astana being the capital of Kazakhstan and their municipalities. Major pollutants were recently detected in soils and water: (1) contaminants from sulfide non-ferrous ores processing: SO_4, NO_3, NH_4, Cu, Pb, Zn, Cd, Tl, Se, Hg, Sb, As, and also a change in pH, (2) complex components, due to processing of rare metal ores, Be and Li, (3) complex detrimental substances: SO_4, Cl, NH_4, NO_3, F, Li, Be, Th, and U with high alpha and beta radioactivity and a change in pH.

Downstream the Irtysh River, the Semipalatinsk Test Site (STS) is situated, where more than 18,000 t of radioactive waste with a total activity of about 1,300,000 Ci, are buried. Radionuclides with a total activity over 10,000,000 Ci were accumulated from the nuclear weapons tests in the underground wells, located within 60 km far from Irtysh River. As a result of multiple nuclear explosions (>450) for the period of 1949–1990, U-235 and Eu-152 are registered on a territory of more than 300 km^2. The Chagan River contaminated with radionuclides flows into the man-made Atomic Lake (created by nuclear explosions) and into the Irtysh river.

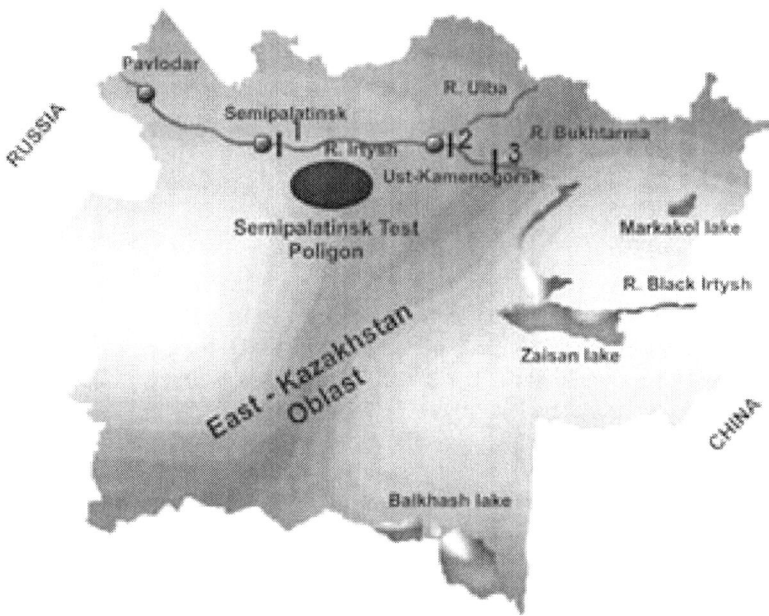

Figure 3. The scheme of the Irtysh River flow in the territory of Kazakhstan Republic 1 – Shulba HES, 2 – Ust-Kamenogorsk HES, 3 – Bukhtarma HES

3. Caucasian Region

On nearly 600,000 m^2 of the highly seismic active Caucasian Region territory, situated between the Black and the Caspian Seas, reside the peoples of Armenia, Azerbaijan, Georgia, North Caucasian part of the Russian Federation, North-east of Turkey and North-west of Iran. The largest Caucasian rivers enter the Caspian and Black Seas.

3.1. ARMENIA

Armenis is located in the basins of Araks and Kura rivers, entering the Caspian Sea, at about 1,800 m mid height above the sea level (Fig. 4). Severe earthquakes, frequent landslides, hail damages, droughts, strong winds and floods are threatening human safety and are causing considerable ecological and economic damages. The 880 MW Armenian Nuclear Power Plant (ANPP), where current protective constructions do not meet the requirements of today, represents a potential man-induced sources of hazard. Besides, the ANPP was designed assuming a maximum grade 7 seismic activity in the region where it is located. In fact, the activity can be of 8–9 grades. In such a case, a nuclear accident may affect a region populated by 20 million people. Other hazardous sites are the cascades of large HES and reservoirs, such as the huge Akhuryan reservoir with the 59 m high dam and the 525 million cubic metres water volume; multiple chemical and mining-chemical productions and their tailing dumps for toxins/radionuclides; as well as gas- and pipelines and power lines. Overall, the above situation is also typical for Georgia and Azerbaijan.

3.2. AZERBAIJAN

In Azerbaijan is located a completely full "mortuary" which does not meet current operational requirements. Multiple sites containing chemical hazards and radioactive substances with intensity of about 500–600 µR/h have been formed in the oil fields (Fig. 5). Radium waters extracted jointly with oil create extra natural contamination. The country has 785 km of coastline with the Caspian Sea. About 100,000 operating radiation-, chemical-, biological-, oil-, and explosive-hazard productions and technologies represent a major threat. During the period of 1992–1998, more than 100 emergencies of man-induced nature were registered. They were followed by explosions and fires, bringing irreversible ecological damage to the entire region, including the Caspian Sea. Furthermore, the majority of productions have sizeable reserves of chemical-hazard substances, such as chlorine, ammonia, hydrochloric acid, and others, and there are up to several thousands of tons of toxins stored at single sites. Azerbaijan suffers from huge extra losses due to flooding of dense populated coastal areas as a result of the water level rise (up to 3 m) in the Caspian Sea. For the period of 1978–1995, the total damage from such flooding amounted to 12 billion USD.

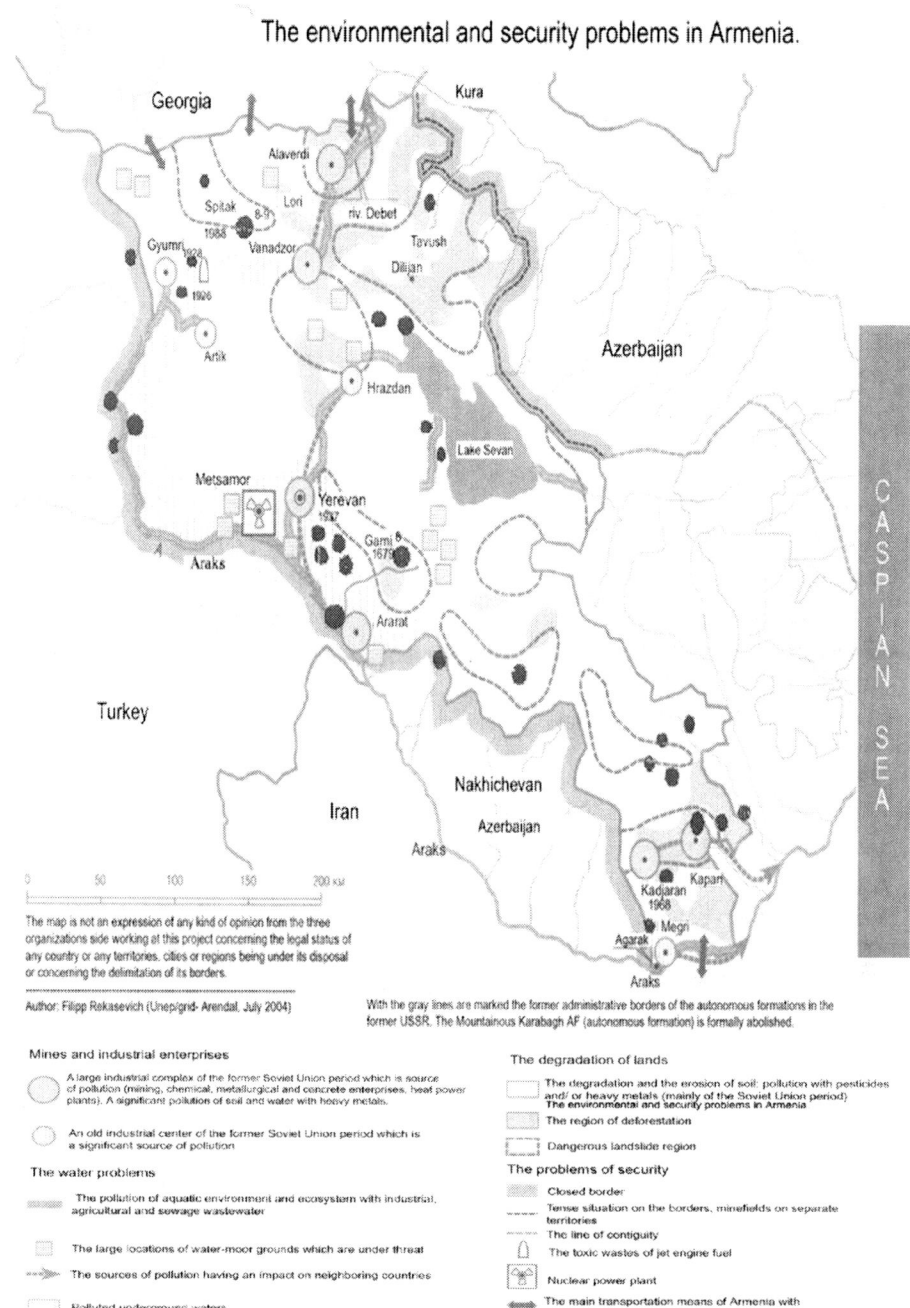

Figure 4. The map of risks in Armenia

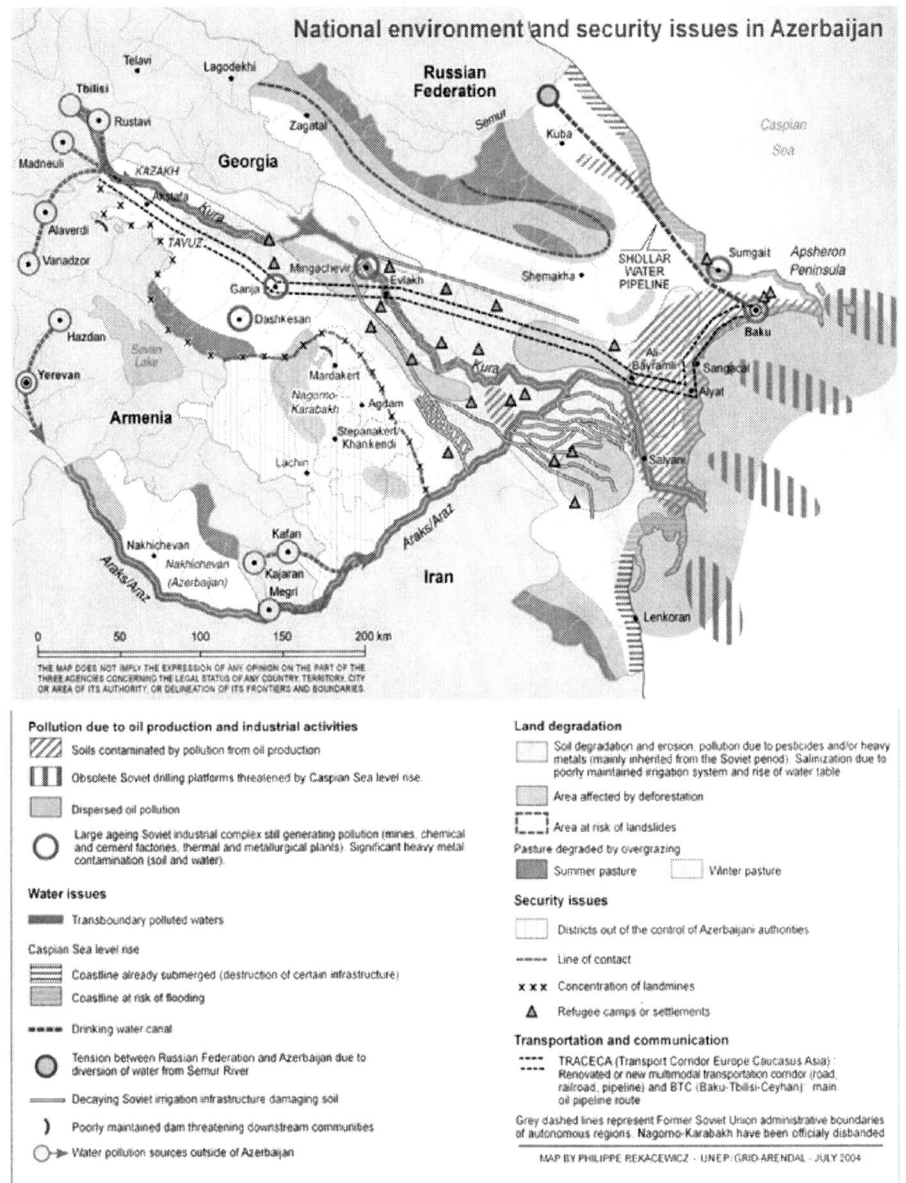

Figure 5. The map of risks in Azerbaijan

At present the Kura River, the largest in the Caucasus, is strongly regulated by the cascade of operating large hydroelectric stations (HES), such as the Mingechauri HES (83 m high dam and volume of 16 billion cubic metres water reservoir), the

Shamkir HES (70 m high dam and volume of 2.7 billion cubic metres), and the Enikend HES (36 m high dam and volume of 158 million cubic metres). All dams of the reservoir are in the emergency state. Above 150 different kinds of chemical substances are registered in the Kura-River waters. Average annual concentration of water contamination exceeds the Maximum Permissible Concentration (MPC) with a factor of 2, and up to a factor of 100 during emergencies; oil product and phenol concentrations exceed MPCs with a factor of 10 to 100.

About 3,800 km of the oil- and above 10,000 km of the field pipelines are in operation at present, while the deterioration of oil-and pipeline affects about 60% of the system. The areas of contaminated sites in Apsheron includes more than 30,000 ha with a radiation level up to 1,200 µR/h. In case of a dam brake in any of the HES, multiple products from different productions and their waste will be spread by a flooding wave over vast lands, including the Caspian Sea. At the same time, the contaminants accumulated in great quantities in sediments and coastal soils for decades will rise from the bottom of the Kura-River and its tributaries and can be spread over large distances.

3.3. GEORGIA

Among the three major HES in Georgia, namely the Zhinval, Shaor and Inguri HESs, the latter represents the major hazard. It has one of the world's largest arch dams, i.e. 271 m height and about 12, 1 billion cubic metre volume of reservoir (Fig. 6). A dam breakage will create a flooding wave with a front height of the wave of ~25 m, which is able both to wash off vast lands of Georgia and to reach Turkey. Possible accidents at the two HESs (of 700 million cubic metres and 200 million cubic metres, respectively), being under construction on the Khudoni and Pari cascades upstream the Inguri-River, will aggravate the situation, as they may double the front height of the wave. Dam breakage at the Shaor HES will create flooding of the cities of Kutaisi and Poti, and flooding from the Zhinval HES will cover the Tbilisi – the capital of Georgia as well as regions of Georgia and Azerbaijan. An accident at the Madneuli mining facility that uses cyanide technologies for extraction will contaminate the Kura-River and the Caspian Sea. Accidents at the Urev and Kvaiss mining facilities will contaminate, apart from major lands, the Black Sea basin with extraction products such as arsenic and antimony. Wash-off or leaching from the tailing dump sites designed for toxic chemical compounds/radionuclides (Lilo, Eastern Georgia), as well as from high toxic rocket fuel repositories in Western Georgia will transport contaminants into the basin of the rivers of Kura and Rioni, and further into the Caspian Sea and the Black Sea. Large ecological catastrophes are possible at the oil and gas terminals in the cities of Supsa and Batumi, as well as in sites where major oil pipelines (Baku-Jeikhan, Baku-Supsa, Novorossiysk-Abkhazia-Turkey), and gas pipelines (Baku-Erzrum, Vladikavkaz-Tbilisi-Yerevan) are operated, partly lying at the bottom of the Black Sea.

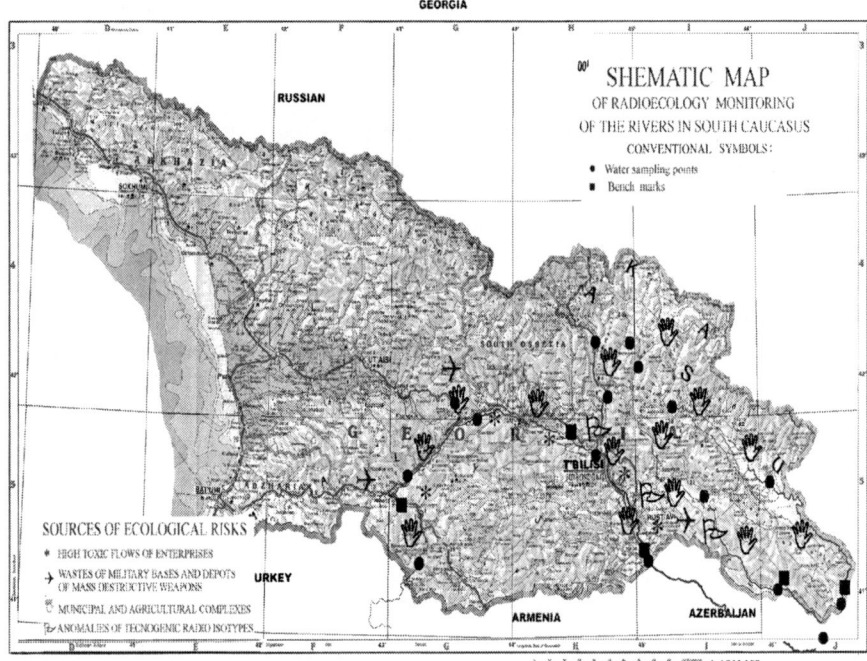

Figure 6. The map of risks in Georgia

Therefore, for all noticed regions, we may single out the following typical factors, that significantly increase the risk for natural and man-induced catastrophes to occur:

- All the regions are located in the mountain areas that have high seismic activity, i.e. level from 5 to 9 grades by the Richter scale.
- The largest mountain rivers have cascades of powerful HES with sizeable water reservoirs and huge high dams (>100 m).
- Within the densely populated regions, there is a series of mines for extraction of metals and minerals, as well as industrial facilities and plants, including power nuclear plant with U-tailing dump sites and burrows of varied pollutants. The facilities use different radioactive, toxic and poisonous substances in their technologies.
- The man-induced activity in the regions under review increases the probabilities for occurrence of not only severe man-induced catastrophes, but also natural ones.
- An especially grave situation has been created on transboundary lands of the states, due to the lack of common ecological and geochemical monitoring systems, that increases political and economic tensions between the countries generating negative migration processes.

- Risks and ecological-economic damages from catastrophes are not only regional, but also global by nature, since they entail contamination of vast lands, the basins of the Black, Caspian and Kara Seas, the Atlantic and Arctic Oceans and consequently the entire World Ocean.

The opportunity to perform deliberate attacks of terrorists with the use of explosives may cause man-induced catastrophes and stimulate natural calamities (earthquakes, mudflows, landslips, etc.). It is easy to implement planned attacks of terrorists there, due to available borders with current centres of international terrorism located in Chechnya, Afghanistan and some others. Especially great is the hazard for new independent states, where the system of safety, boundary and customs control, that of strict visa control and other state safety measures have not yet been formed. Consequences from attacks of terrorists in the regions can be followed by major human and huge material losses, and extremely negative irreversible environmental effects of global scale. Some special space technologies will be used in all regions for continuous observations and control of the most dangerous large objects [10–12].

4. The Calculation of the Total Limited Losses

Our method is the following: we consider a common case of any object for a fixed time interval under the following assumptions: (1) at initial state the object is in normal (non-accidental) exploitation/operation/situation; (2) the different kinds of accidents which may occur as noticed i = 2, 3,... m, where m is the total number of possible accidents (m = 1 corresponds to the normal regime); (3) every accident may create different kinds of losses. Assume that j is the kind of loss with a_j value, then j = 1, 2, ...n, where n is the total number of possible kinds of losses; (4) realization of i accident creates the loss of j kind with P_{ij} probability, thus the matrix of loss probabilities is determined. Then, the total vector of limited losses $\vec{a}_{lim}$ may be determined by the following equation:

$$\vec{a}_{lim} = P(1)\vec{a}_{1n} + \sum_{i=2}^{m} \hat{P}_{ij} \vec{a}_j \qquad (1)$$

where P(1) is the probability of loss formation under normal exploitation; $\vec{a}_{1n}$ is the vector of limited loss under regular exploitation. $P_{ij}a_j$ coordinate vector value in sum is equal to the loss value of j kind under realization of i type of an accident.

Under absence of accidents, the second term in the right part of (1) equals zero and then $\vec{a}_{lim}$ total vector of limited loss is determined by the first part of (1):

$$\vec{a}_{lim}n = P(1)\vec{a}_{1n} \qquad (2)$$

The main problem in this calculation is the determination of the loss probability matrix. As one of the possible methods, we propose to use the method of expertise estimates. The calculations of the total losses includes the following main moments: at first for every object we have to point out and develop the classification of a major possible accidents. For example, in the case of a HES disaster we have to take into account the following possible types of accidents:

1. Total damage or breakage of one or more HES dams
2. Partial damage of HES dam
3. Destruction of water locks
4. Stopping of HES turbines.

Let us consider the most dangerous accident, investigate the extreme cases of the worst possible catastrophes and analyze the possible scenarios connected to the damages of two HES, located upriver of Irtysh near Ust-Kamenogorsk city (Fig. 3)

1. Bukhtarma hydro electric stations (HES)
2. Ust-Kamenogorsk HES
3. Both HES simultaneously.

Here, we have to take into account that the total damage of Bukhtarma HES dam with the height of ~100 m will probably stimulate the total damage of Ust-Kamenogorsk HES with the height of ~40 m. Then, it is necessary to evaluate the parameters of catastrophic submergence in every scenario:

- Maximum possible height and speed of break-through wave
- Estimated time of wave crest and the front of wave crest coming into town territory
- Boundaries of possible submergence zone in the vicinity
- Maximum depth of submergence for every definite locality ant time of its submergence
- To point out all main objects that will be flooded.

For these estimates and calculations we shall use the computer modelling, taking into account the real profiles of the local earth crust and mountains valley (including its rock and soil materials), the HES with all connected water reservoirs, such as lakes and rivers, and other natural objects.

In case of a possible HES disaster near U- tailing storage aites (Fig. 2), our analysis will include the following:

1. Analysis of possible scenarios leading to pollutant migration from tailing storage
   - Constant pollutant migration without damage of tailing storage dams
   - Similar migration with partial damage of tailing storage dams, for example, under landslide or earth flow
   - Pollutant migration under complete damages of dams, for example in result of an earthquake
   - Pollutant migration in result of
2. Partial flow
3. Total flooding.

Under realization of last two scenarios, it is possible two follow the development of the catastrophic situation:

1. All tailings are washed off with river during few days
2. All tailings are washed off with river instantly.

The last situation is the most extreme and dangerous, because it will cause the maximum pollution with maximum losses both for environment and population.

For all cases it is necessary to take into account the following types of possible losses:

1. Caused by injuries to the population (mortality) or harm to population health
2. Caused by pollution of wide scale territories with subsequent losses associated with forest, agricultural and fish industries
3. From heavy pollution of buildings and constructions
4. From the pollutant migration into basins of the largest rivers.

In risk evaluations it is necessary to take into account the possible chemical and nuclear reactions and transformations of pollutants in soil, water and air. For example, transport calculations will be done for the decay chain $^{238}U > {}^{234}U > {}^{230}Th > {}^{226}Ra$.

Project objectives are as follows:

- Selection for each country a site for which the risk of one or several catastrophes is maximum and where the damage is the greatest
- Development of scenarios for implementing possible catastrophes for the site selected
- Estimation of risks and possible ecological and economic damages at different scenarios of catastrophe development
- Suggestion of some recommendations on risk reduction and actions to eliminate the effects of accidents or catastrophes.

5. Conclusion

For all presented regions we singled out, the following typical factors that increase significantly a risk of natural and man-induced catastrophes to occur:

1. These regions are located in the mountain lands with the high seismic level (5–9 grades by Richter scale).
2. The largest mountain rivers have cascades of powerful hydroelectric stations with their sizeable reservoirs and huge high dams (>100m).
3. In the regions which are densely populated, there are plenty of mines for extraction of metals/minerals, industrial facilities and plants with U-tailing dumps and burrows of varied pollutants with using the different radioactive, toxic and poisonous substances in their technologies.
4. The man-induced activity here increases probabilities for occurrence of not only severe man-induced catastrophes, but also natural ones.
5. An especially grave situation has been created on trans boundary lands of the states, due to the lack of common ecological and geochemical monitoring systems, that increasing political and economic tension between the countries and generating of negative migration processes.
6. Risks and ecological-economic damages from catastrophes are not only regional, but also global by nature, since they entail contamination of vast lands, the basins of the Black, Caspian and Kara Seas, that of the Arctic Ocean and, consequently, the entire World Ocean.

7. Opportunity to perform deliberate terrorist attacks with the use of explosives, that are able to cause man-induced catastrophes and stimulate natural calamities (earthquakes, mudflows, landslips, etc.). It is easier to perform attacks by terrorists there, due to the intersection of main lines, available border with current centres of international terrorism, located in Chechnya, Afghanistan and some other places. Especially great is the hazard for new independent states, where the system of safety, boundary and customs control, that of strict visa control and other state safety measures have not yet been formed. Consequences from attacks of terrorists in the regions will be followed by major human and huge material losses, and extremely negative irreversible environmental effects of global scale.

The result of the Program will be used in the following way:

1. Evaluation of the risk values from possible natural or man-made catastrophes for the most dangerous objects for the development of a methodology/strategy to regulate and manage risks in emergencies
2. Mapping risk associated with various lands
3. Developing a common system for emergency prevention/elimination. To formulate the preventive countermeasures and to estimate their efficiency using the resource parameters for decreasing risks values, their preventions and softening of their responses.

The obtained results will have a universal character and may be used for analysis of similar objects and situations in other countries. This Program will promote the concept of substantial development with growth, economical cooperation and stability, decreasing political stress not only for the countries participating, but also at global scale for all countries, located at the continent.

Reference

1. Valyaev A.N. and Yanushkevich. (2004) Using of Acoustic Techniques for Detection of Explosives in Gas, Liquid and Solid Mediums. *NATO Science Series II "Detection of Bulk Explosives: Advanced Techniques Against Terrorism"* (Mathematics, Physics and Chemistry) 138, pp. 175–183 and June 16–21, 2003, Kluwer, The Netherlands. Proc. of NATO Advanced Research Workshop, St. Petersburg, Russia.
2. Stepanets O.V., Borisov A.M., Ligaev A.R., Vladimirov M.K. and Valyaev A.N. (2005) Estimation of Parameters of Radioactive Environmental Contamination in Places of a Burial Objects in Shallow Bays of Archipelago Novaya Zemlya on Data 2002–2004 *Abstracts of 5-th International Conference "Nuclear and Radiation Physics"* September 26–29, 2005, Almaty, Kazakhstan, pp. 463–464.
3. Kazakov S.V., Utkin S.S., Linge I.I. and Valyaev A.N. (2005) Categorization of Water Media and Water Bodies by the Level of Radioactive Contamination. *Abstracts of International Conference "Radioactivity After Nuclear Explosions and Accidents."* December 5–6, 2005, Moscow, Publ. House: St. Peterburg, GIDROMETIZDAT, pp. IV-20–IV-21.
4. Valyaev A.N., Kazakov S.V. and Petrov V.A. (2004) Estimations of Risks, Ecological and Economical Losses Under Catastrophic Damages of Hydro Stations in East Region of Kazakhstan Republic in *Proc. of Intern. Symposium "Complex Safety of Russia – Investigations, Management, Experience."* May 26–27, 2004, Moscow, Publ. House: Informizdatcenter, pp. 348–353. (in Russian)
5. Valyaev A.N., Kazakov S.V., Aitmatov I.T. and Aitmatova D.T. (2004) Estimates of Risks, Ecological and Economical Losses from the Emergence of Natural and Man–made Catastrophes in Uranium Tailing Storages in Tien-Shyan Mountains in *Proc. of Intern. Symposium "Complex Safety of Russia –*

Investigations, Management, Experience." May 26–27, 2004, Moscow, Publ. Housse: Informizdatcenter, pp. 353–358. (in Russian)

6. Kazakov S.V., Valyaev A.N. and Petrov V.A. (2004) Estimations of Risks, Ecological and Economical Losses Under Catastrophic Damages of Hydro Stations in East Region of Kazakhstan Republic in *Proc. of Intern. Russian-Kazakhstan Scientific – Applied Conference* (October 5–6, Ust-Kamenogorsk, Kazakhstan) 2004, Part 1, pp. 330–333.

7. Valyaev A.N., Kazakov S.V., Aitmatov I.T. and Aitmatova D.T. (2004) Earth from Space – Some Actual Problems and Their Possible Decisions with Using of Space Monitoring in *Proc. of Intern. Russian-Kazakhstan Scientific – Applied Conference* (October 5–6 2004, Ust-Kamenogorsk, Kazakhstan) Part 3, pp. 196–199.

8. Passell H.D., Barber D.S., Kadyrzhanov K.K., Solodukhin V.P., Chernykh E.E., Arutyunyan R.V., Valyaev A.N., Kadik A.A., Stepanets O.V., Alizade A.A., Guliev I.S., Mamedov R.F., Nadareishvili K.S., Chkhartishvili A.G., Tsitskishvili M.S., Chubaryan E.V., Gevorrgyan R.G. and Pysykulyan K.A. (2005) The International Project of Radiation and Hydrochemical Investigation and Monitoring of General Caspian Rivers. *Abstracts of 5-th International Conference "Nuclear and Radiation Physics."* September 26–29, 2005, Almaty, Kazakhstan, pp. 487–489.

9. Tsitskishvili M.S., Kordzakhia G., Valyaev A.N., Kazakov S.V., Tsitskishvili N., Aitmatov I.T. and Petrov V.A. (2005) Insurance of Risk Assessment and Protection from Distant Transportation and Fall out of Pollutants under Large Anthropogenic Damages on Nuclear Power Stations due to Mountainous Regional Peculiarities. *Abstracts of 5-th International Conference "Nuclear and Radiation Physics."* September 26–29, 2005, Almaty, Kazakhstan, pp. 460–461.

10. Valyaev A.N., Kazakov S.V., Aitmatov I.T. and Aitmatova D.T. (2003) Problems of Ecologic Safety under Displacement of Rock Stones, controlled from Space, at Ecological Dangerous Regions of Tien-Shyan Mountains in *Proc. the First Intern. Conference "Earth from Space- the Most effective Solutions."* November 26–28, 2003 http://www.transparentworld.ru/conference/presentations/ operative.htm tyan_ shyan_prsnt.zip.

11. Valyaev A.N. and Kazakov S.V. (2003) Earth from Space – Some Actual Problems and Their Possible Decisions with Using of Space Monitoring in *Proc. the First Intern. Conference "Earth from Space- the Most Effective Solutions."* November 26–28, 2003 http://www.transparentworld.ru/conference/ presentations/operative.htm.

12. Valyaev A.N., Kazakov S.V., Stepanets O.V., Malyshkov Yu.P., Solodukhin V.P., Aitmatov I.T. and Aitmatova D.I. (2005) Using of Space Technologies for Monitoring of Large River's Basins and Prediction of Large-Scale Natural and Man-Induced Catastrophes in Ecology-Hazard Regions of Central Asia and Caucasus. in *Proc. the Second Intern. Conference "Earth from Space-the Most Effective Solutions" Section "Space Monitoring in Problems of Management of Territories."* http://www.tran- sparentworld.ru/conference/2005/thesis.pdf .

CHAPTER 14

DISTRIBUTION OF NATURAL AND TECHNOGENIC RADIOACTIVITY IN SOIL SAMPLES FROM FOOTHILL AND MOUNTAIN AREAS IN CENTRAL TAJIKISTAN

A.A. JURAEV[1], AN.A. DZURAEV[1],
T. DAVLATSHOEV[1], H.D. PASSELL[2]
[1] *Physical-Technical Institute of the Academy of Sciences of Republic of Tajikistan (PhTI)*
[2] *Geosciences and Environment Center Sandia National Laboratories, Cooperative, Monitoring Center, USA*

Abstract: Within the framework of the multinational "Navruz" project (Kazakhstan, Kyrgyzstan, USA, Tajikistan and Uzbekistan), funded by ISTC (T-1000 and T-1163), the Physical-Technical Institute of Academy of Sciences of Republic of Tajikistan, has been involved in the research on identifying radioactive contamination of transboundary rivers and their basic inflows. Samples of water, water suspensions, sediments, water vegetation and soils have been collected and analyzed. In the present paper results on the distribution of uranium and thorium series radioactive isotopes, K-40 as well as Cs-137 in the Hissar and Zarafshan mountain ranges slopes are given. The results from 2005 showed that the Cs-137 concentration in soils varied from several tens up to hundreds Becquerel per kilogram.

Keywords: uranium, thorium, Cs-137, soils, Tajikistan

1. Introduction

Tajikistan is a mountainous country. Northern and central parts of Tajikistan are divided by a mountain system consisting of Hissar, Zerafshan and Turkistan ranges. This mountain system has an interesting feature; these ranges cross the territory of Tajikistan practically from east to west, beginning on the east side and continuing as the Alay range in Kyrgyzstan, and ending at the west side in the Samarkand, Surkhan-Darya and Kashkadarya areas of Uzbekistan. Due to the rather large heights of these mountains (the peaks reach 5 and more than kilometers) these mountains have areas which are covered by snow and glaciers. This high-altitude and low-temperature area represents a barrier for moving air masses both from the Arctic Ocean and for the southern monsoons from the Indian Ocean, creating an intermountain area with unique climatic conditions. For example, the Hissar Range Mountains are practically completely blocking the transport from the north of the powerful dust storms, which are formed in region and known under the name "Afghan". The area is also known as topic a beauty of the mountainous Tajikistan.

The Physical-Technical Institute of Academy of Sciences of Republic of Tajikistan, named after S.U. Umarov, has been a participant in the multinational project "Navruz" (Kazakhstan, Kyrgyzstan, USA, Tajikistan and Uzbekistan), funded by USA via ISTC, since 1999. The objective of the project is to characterize the radioecological conditions of transboundary rivers and their basic inflows. The research is based on the collection of a series of samples; water, water suspensions, sediments, water vegetation and soils [1, 2]. Already in the first samples analyzed at the Sandia National Laboratories, USA, traces of artificial radioactivity such as caesium-137 were identified [3], as shown in Fig. 1. The results also formed the basis for drawing a conclusion; that in a mountain part of Tajikistan enrichment of uranium had been taking place. To identify the sources and the reasons of the obtained results, we paid attention to results from the air monitoring radioactivity control at the Institute of Nuclear Physics (INPh) of the Academy of Sciences of Republic of Uzbekistan (Fig. 1).

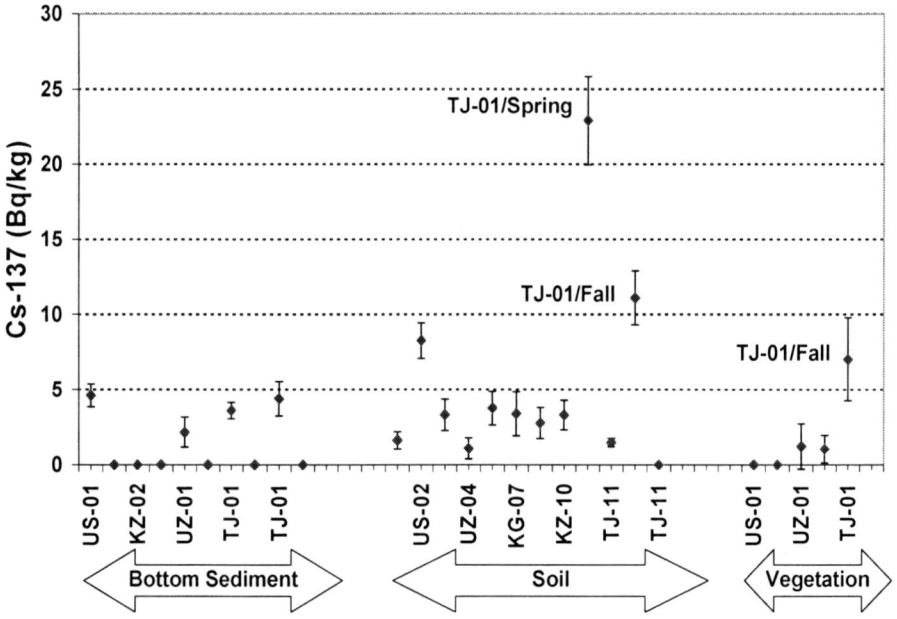

Figure 1. Concentration of Cs-137 in the samples collected in 2000

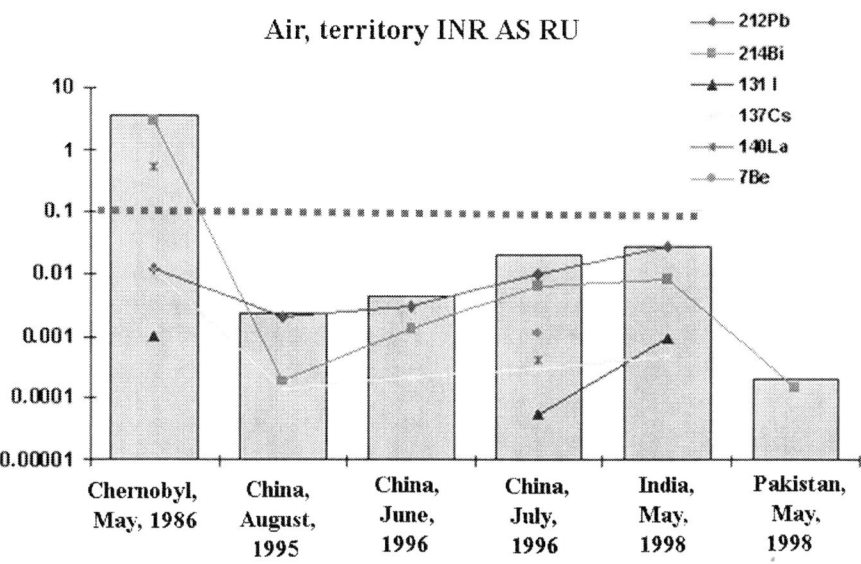

Figure 2. Air monitoring of radioactivity at INPh AS RU

Figure 3. Site of the basic objects capable to influence radiating conditions in republic Tajikistan

The results from the air monitoring control, provided by the colleagues from Uzbekistan, showed that the increase in air radioactivity at INPh in Uzbekistan correlated well with global fallout from nuclear weapons tests in neighbour countries and with fallout from large nuclear accidents (Fig. 2). Fallout of radioactive pollution from the nuclear accident at Chernobyl Nuclear Power Station is also clearly identified. Radioactivity from nuclear tests, announced as underground tests, can be explained by a large output to the surface of radioactive xenon and krypton isotopes, having maximal output (yield) in a fission reaction and by bad "tightness" of used devices.

In April, 2005 in Vienna (IAEA), at the international seminar "Radiology and Application", we showed a map of regional features [4]. On this map (Fig. 3) the green point corresponds to the mountain systems included in the present work, and the red points mark the tentative locations of the nuclear weapon testing areas. Looking on this map we should be concerned about the situation in the Islamic Republic of Iran, not only because the current events can result in the development and consequently to testing of nuclear weapons. We are also concerned about the development of a nuclear centre and an increase in risks associated with a nuclear accident connected to the peaceful use of an atomic energy.

2. Methods

Surface soil layers (1–2 cm) were collected from plots of 40 × 40 m^2. In laboratory, large fragments of inclusions and roots were removed. The remains were sieved through a sieve of 1 mm, packed, and stored for future analysis.

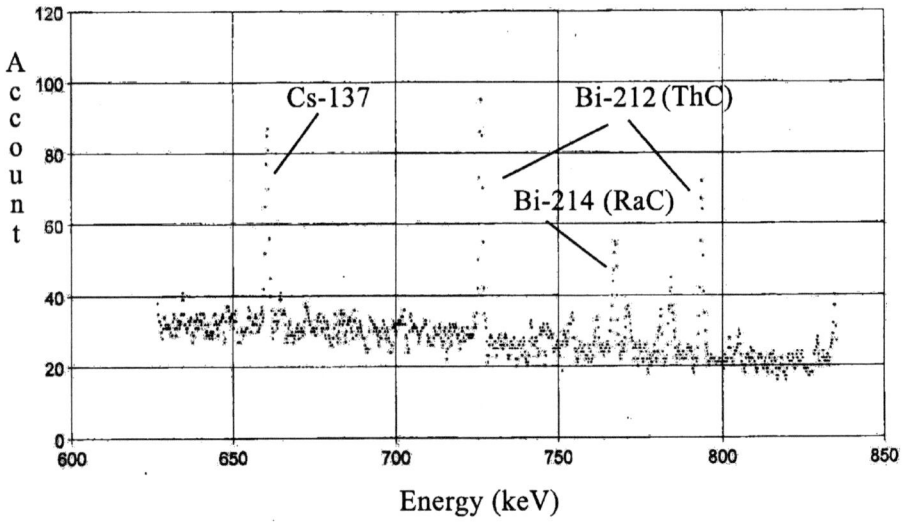

Figure 4. Typical gamma spectrum for soil samples investigated

Gamma-radiometric analysis was carried out using a Canberra detector with relative efficiency 10%. The measurements were carried out in a lead shielded box with 5 cm wall thickness and covered inside by 2 mm cadmium layer. The internal surface of the box was decontaminated before each measurement session. Throughout the measurements, the sample was contained in Marinelly beakers having a total volume 500 cm^3.

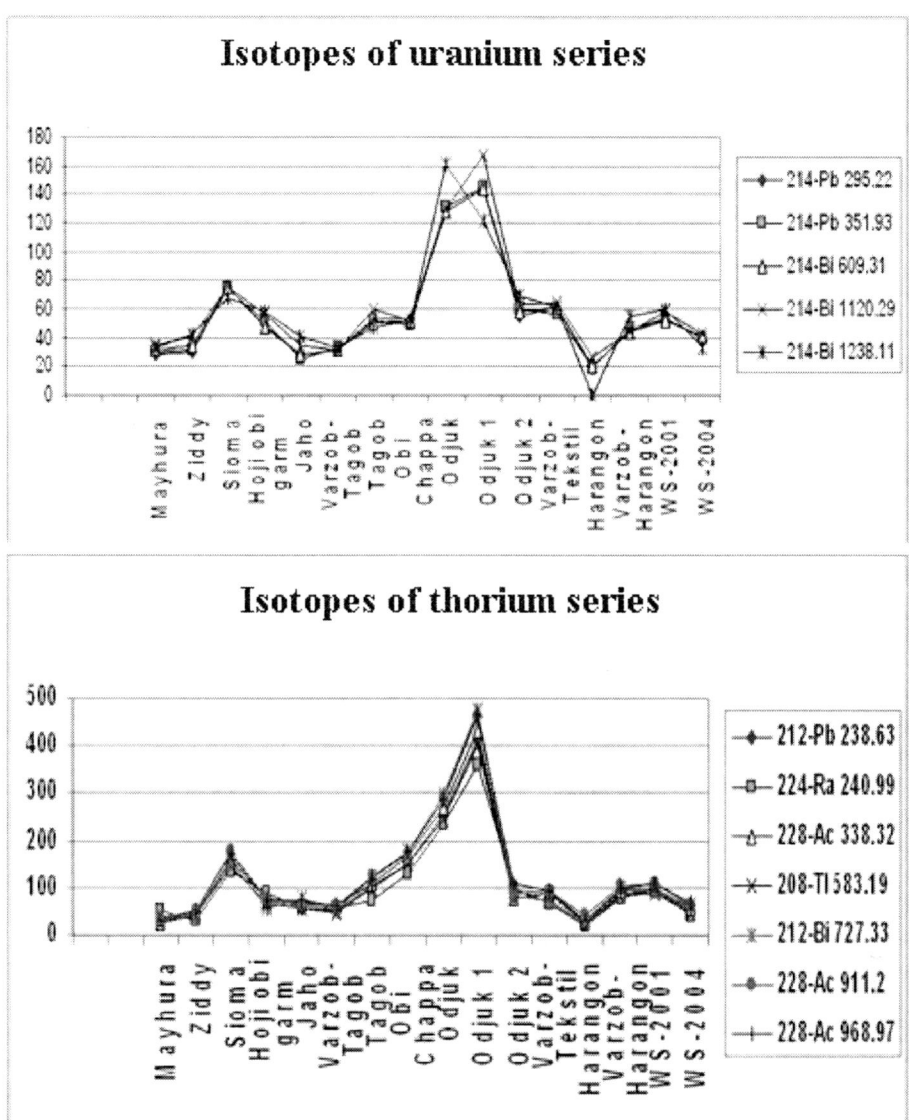

Figure 5. Accuracy of the gamma measurements for the determination of radioactive isotopes in uranium and thorium series

The measurement time for the samples and standards was chosen to be 6 h, and the measurement time for the background level (in the beginning and end of a series of measurements) was 24 h. One of the spectra is shown in Fig. 4. It is visible, that the used system has the necessary properties (efficiency, resolution) for the present application. The accuracy of the measurement method was determined, based on 21 samples and 5 gamma-lines from isotopes of uranium series and on 7 lines from isotopes of thorium series [5, 6], as illustrated in Fig. 5.

3. Experimental Results and Discussion

The survey results based on 55 sampling points are presented. For each sampling point the absolute coordinates (longitude and latitude) have been defined. The data are represented as numbers of degrees (decimal of degrees. Such representation has allowed us to obtain spatial information on the distribution of radionuclides in the samples investigated. Thus, we have divided the sample material into four parts:

- The foothills of Hissar valley, from west on east
- The Hissar range, from west on east
- The Turkestan and Zerafshan ranges, from west on east
- The gorge of the Varzob River, from north on the south.

The western border is the Surkhan-Darya and Samarkand areas in the Republic of Uzbekistan, and the east border is the Batken area in Republic Kyrgyzstan. For each allocated site, results are given for the specific activity concentration (Bq per kg) of elements in the uranium and the thorium series. Detailed results are given in Tables I–IV and in Figs. 6–9.

4. Conclusions

In all samples, the concentrations of Pb-210 are easy to determine, and the Pb-10 concentration rises with a factor of 10 in some cases. This has been marked as excess Pb-210 in comparison with the equilibrium concentrations of other isotopes in the uranium series. This excess, apparently, is caused by radon deposition on the ground formed by nearby rocks rich in granites and marbles and with high concentration of uranium. The radiological risks of such a situation is related to this radioactive lead, which via a juicy grass at the foothill area can be transferred into the human body. It is therefore necessary to carry out research on the radon distribution and to establish an appropriate control service in places for people residing in the mountain and foothill areas.

Attention is also put on the distribution of the uranium–thorium series isotopes and on the distributions of K-40, and the correlation between these radionuclides. The concentration of K-40 exceeds the average level in areas where the increase of the concentration of uranium and thorium also is notably. This fact indicates that the same geochemical processes are responsible for the increase of the concentration of uranium as well as for the increase of the potassium concentration.

The analysis of the distributions of the technogenic isotope Cs-137 shows that the concentration in the investigated samples changes from several Becquerel per kg to hundreds Becquerel per kg. The results for enriched Cs-137 correspond to the concentrations of diffused Cs-137 as a result of nuclear weapons tests and accidents within the region of Central Asia. Thus in the foothill areas, the maximum concentration of Cs-137 was observed in the western part, close to the Karatag River, being an inflow to the Surkhandaria River. Elevated Cs-137 concentrations are observed farther; close to the central part of the Hissar range (source of the Kafirnigan River), crosses into an area being the source of Varzob River, then into the Fandaria gorge where it crosses the Zaravshan range and passes through the Turkestan range in the mountainous area of the Shakhristan Pass. It is possible to explain such distributions of the Cs-137 contamination by wind transport in the beginning of a summer, where the plume has a primary direction from the south on north.

References

1. Djuraev, A.A. et al. (2002) Radioecological Situation in river Basins of Central Asia, Syrdaria and Amudaria: Results of the international project NAVRUZ, Advanced Research Workshop. "Environmental protection Against Radioactive Pollution" Abstracts, September, 16–19, Almati, pp 84–86.
2. Djuraev, A.A. et al. (2002) "The "Navruz" Project, Monitoring for radionuclides and metals in Central Asia transboundary rivers", End of year one reports, Sandian report, Sand 2002–2916, September, 84 pp.
3. Mohagheghi, A.H. (2001) "A Cooperative Transboundary Water Monitoring Project in Central Asia, The "NAVRUZ" Experiment", DE-AC04-94AL86000, http://www.sandia.gov.
4. Djuraev, A.A, Djuraev, A.A. Jr. and Davlatshoev, T. Physical-technical Institute Academy of Sciences (Republic of Tajikistan), Mamajanov, J. Institute of Geology Academy of Sciences (Republic of Tajikistan), Passell, Howard D. Sandia National Laboratories, DOE (US), (2005), "Radioactive pollution in the Varzob River Gorge", report on "Second International Workshop on Radiological Sciences and Applications (IWRSA)", Vienna, Austria, March 16–18.
5. Djuraev, A.A. and Djuraev, A.A. Jr. (2004) The Choice of Points of Supervision of Radioecological Monitoring of the Rivers of Tajikistan, ISTC Report, Dushanbe, pp. 12.
6. Dzhuraev, A.A. et al. (2004) Allocation of Connatural and Technogenic Radionuclides in Bottom Sediments of the Rivers of Varzob Gorge, Thesis of OSCE Conference "Monitoring of migration and accumulation of radionuclides in component of natural ecosystems" Dushanbe.

Appendix

TABLE I. Distribution of natural and technogenic radioactivity in soil samples of soil from foothills of Hissar valley, from west on east

Sample	Lat	Lon	River	^{234}Th	δ	^{210}Pb	δ	^{228}Ac	δ	^{212}Pb	δ	^{40}K	δ	^{137}Cs	δ
TJ-58	38.567	68.317	Karatag	15.6	2.3	129.9	19.5	36.1	5.4	34.3	5.1	571.4	40.0	21.74	3.26
TJ-64	38.550	68.383	Gissar channel	19.8	2.9	44.6	6.7	43.8	6.5	33.7	5.0	610.7	42.7	0.00	0.00
TJ-59	38.583	68.550	Khonako	21.7	3.2	50.2	7.5	48.7	7.3	40.2	6.0	696.4	48.7	0.69	0.10
TJ-57	38.450	68.733	Shirkent	14.6	2.2	68.6	10.3	61.3	9.2	47.6	7.1	724.9	50.7	3.97	0.59
TJ-60	38.583	68.750	Luchob	21.9	3.2	52.2	7.8	41.0	6.1	31.4	4.7	489.2	34.2	2.31	0.35
TJ-63	38.567	68.750	Gissar channel	16.4	2.4	73.7	11.1	46.5	6.9	40.3	6.0	596.4	41.7	4.87	0.73
TJ-61	38.567	68.767	Varzob	11.7	1.7	80.5	12.1	36.5	5.4	34.0	5.1	496.4	34.7	2.97	0.45
TJ-62	38.567	68.767	Gissar channel	25.6	3.8	24.3	3.6	42.0	6.3	36.6	5.4	660.7	46.2	1.32	0.20
TJ-02	38.517	68.767	Varzob	22.90	2.29	77.0	7.7	79.0	7.9	74.3	7.4	833.8	83.4	3.90	0.39
TJ-02	38.517	68.767	Varzob	41.96	5.29	71.4	14.3	87.5	4.1	67.4	0.92	967.1	17.5	3.14	0.37
TJ-02	38.517	68.767	Varzob	40.77	4.87	59.1	15.7	88.4	3.7	70.1	0.9	916.7	16.8	4.48	0.70
TJ-02	38.517	68.767	Varzob	11.89	1.78	15.4	2.3	34.6	5.2	34.1	5.1	435.7	30.5	1.48	0.22

TABLE II. Distribution of natural and technogenic radioactivity in soil samples from the Hissar range, from west on east

Sample ID	Lat	Lon	River	Th-234	δ	Pb-210	δ	Ac-228	δ	Pb-212	δ	K-40	δ	Cs-137	δ
TJ-70	38.650	68.567	Рогова	26.3	4.2	24.1	13.3	45.1	2.9	37.6	0.7	681.8	13.8	0.0	0.3
TJ-71	38.650	68.567	Хонако	20.3	4.1	32.3	13.2	46.4	3.1	39.5	0.7	803.4	15.2	1.4	0.3
TJ-68	38.683	68.367	Сабургун	12.0	4.4	31.5	12.5	28.6	2.6	27.6	0.6	694.4	13.8	0.9	0.5
TJ-69	38.683	68.367	Карагаг	26.9	4.4	36.6	11.3	43.9	3.0	35.1	0.7	719.5	14.3	0.0	0.0
TJ-66	38.717	69.317	Сардаи Миена	90.6	6.3	169.3	21.0	207.8	5.5	169.8	1.4	1,306.9	21.7	18.6	0.9
TJ-67	38.783	69.317	Сарбо	70.2	7.0	183.7	20.8	174.4	5.6	140.4	1.3	1,365.6	21.7	40.9	1.0
TJ-56	39.167	70.867	Navobod	34.7	5.2	74.5	11.2	61.0	9.2	50.5	7.6	817.8	57.2	3.4	0.5
TJ-55	39.183	71.200	Hoit	19.0	2.9	0.0	0.0	43.5	6.5	35.7	5.4	549.9	38.5	0.0	0.0
TJ-52	39.267	71.367	Kizilsu	19.5	2.9	66.9	10.0	33.9	5.1	25.7	3.9	396.4	27.7	6.4	1.0
TJ-53	39.267	71.367	Muksu	18.5	2.8	35.0	5.2	37.8	5.7	29.3	4.4	367.8	25.7	1.0	0.2
TJ-54	39.267	71.367	Jirgital	16.8	2.5	55.1	8.3	32.2	4.8	27.4	4.1	403.5	28.2	0.9	0.1

TABLE III. Distribution of natural and technogenic radioactivity in soil samples from the Hissar range, from west on east

Sample	Lat	Lon	River	^{234}Th	δ	^{210}Pb	δ	^{228}Ac	δ	^{212}Pb	δ	^{40}K	δ	^{137}Cs	δ
TJ-44	39.500	67.533	Zarafshan	14.2	1.4	41.8	4.2	47.1	4.7	40.7	4.1	708.9	70.9	0.80	0.08
TJ-45	39.483	67.717	Magian Darya	9.3	0.9	20.5	2.1	33.2	3.3	31.7	3.2	569.6	57.0	1.00	0.10
TJ-75	39.450	68.533	Shakhristan 4	33.7	4.9	72.4	15.3	65.7	3.5	54.4	0.8	1,151.7	19.3	13.59	0.81
TJ-72	39.583	68.567	Shakhristan 1	39.5	3.9	81.6	13.7	52.5	3.2	40.2	0.7	686.0	13.8	21.33	0.87
TJ-73	39.583	68.567	Shakhristan 2	31.3	4.4	73.2	14.0	49.8	3.4	43.6	0.8	950.3	17.1	8.57	0.66
TJ-74	39.567	68.600	Shakhristan 3	33.1	4.3	152.4	14.8	42.1	3.1	37.5	0.7	686.0	13.9	59.69	1.18
TJ-90	39.383	68.600	Inflow Zar.	25.6	4.2	27.2	13.1	49.1	3.2	39.6	0.7	908.3	16.8	0.00	0.00
TJ-89	39.383	68.633	Inflow Zar.	18.9	3.8	49.4	12.0	15.3	1.2	29.1	0.7	333.5	9.1	0.00	0.00
TJ-87	39.383	68.750	Inflow Zar.	20.0	4.1	42.7	13.2	54.4	3.2	44.2	0.8	983.8	17.6	10.70	0.71
TJ-86	39.383	68.783	Inflow Zar.	26.0	4.3	74.0	13.2	54.4	3.3	46.6	0.8	1,025.8	18.1	6.38	0.72
TJ-85	39.400	68.850	Inflow Zar.	28.6	4.5	41.4	31.5	64.3	3.7	50.2	0.8	941.9	18.4	8.87	0.77
TJ-84	39.400	68.883	Inflow Zar.	28.0	4.8	172.1	34.0	66.8	3.8	55.9	0.8	1,118.1	19.1	4.18	0.54
TJ-83	39.383	68.933	Inflow Zar.	27.2	4.8	53.5	12.3	67.9	3.8	55.4	0.8	1,218.8	20.5	0.00	0.00
TJ-82	39.400	69.000	Inflow Zar.	41.2	4.7	58.1	15.4	75.9	3.9	60.8	0.9	1,558.6	23.5	2.51	0.63
TJ-81	39.417	69.017	Inflow Zar.	26.9	4.9	137.8	34.6	68.2	3.6	55.0	0.8	1,248.2	20.4	5.81	0.61
TJ-80	39.417	69.083	Inflow Zar.	29.1	4.5	143.4	14.5	50.5	3.2	43.3	0.8	1,042.6	18.2	11.85	0.81
TJ-79	39.417	69.150	Inflow Zar.	26.8	4.7	54.5	5.5	64.3	3.6	50.0	0.8	1,197.8	19.6	3.03	0.62
TJ-78	39.417	69.217	Inflow Zar.	33.8	4.6	57.7	14.6	56.3	3.5	51.7	0.8	1,067.8	19.0	6.13	0.73
TJ-77	39.417	69.267	Inflow Zar.	43.2	4.5	201.7	30.9	64.0	3.6	49.3	0.8	1,067.8	18.5	8.66	0.72
TJ-76	39.417	69.367	Inflow Zar.	32.7	4.7	55.3	12.4	63.6	3.6	46.8	0.8	1,109.7	19.0	7.85	0.67

TABLE IV. Distribution of natural and technogenic radioactivity in soil samples from the gorge of the Varzob River, from north on the south

Sample ID	Lat	Lon	River	^{134}Th	δ	^{210}Pb	δ	^{228}Ac	δ	^{212}Pb	δ	^{40}K	δ	^{137}Cs	δ
TJ-25	38.967	68.750	Sioma	47.8	7.5	476.0	19.8	157.9	5.1	127.4	1.3	1,139.1	19.4	141.1	1.6
TJ-65	38.967	68.750	Varzob	36.6	7.2	311.5	16.4	135.1	4.8	106.2	1.1	1,269.1	20.6	24.9	1.0
TJ-28	38.850	68.850	Tagob	12.3	1.2	46.7	4.7	51.7	5.2	45.4	4.5	637.4	63.7	1.0	0.1
TJ-50	38.833	68.850	Varzob-Tagob	44.2	4.4	103.8	10.4	187.9	18.8	166.4	16.6	773.1	77.3	4.3	0.4
TJ-29	38.800	68.817	Obi Chappa	51.8	5.2	164.6	16.5	186.7	18.7	155.4	15.5	1,044.5	104.5	12.1	1.2
TJ-47	38.700	68.783	Odjuk 2	82.8	8.3	129.9	13.0	340.8	34.1	289.7	29.0	1,351.7	135.2	28.0	2.8
TJ-30	38.683	68.817	Odjuk	18.0	1.8	151.8	15.2	90.0	9.0	75.2	7.5	698.1	69.8	29.5	3.0
TJ-46	38.683	68.800	Odjuk 1	90.2	9.0	128.6	12.9	329.9	33.0	285.3	28.5	1,130.2	113.0	9.4	0.9
TJ-49	38.683	68.767	Varzob-Harangon	26.3	2.6	62.9	6.3	95.8	9.6	81.8	8.2	801.7	80.2	4.4	0.4
TJ-48	38.667	68.767	Harangon	16.3	1.6	79.0	7.9	56.4	5.6	48.1	4.8	741.0	74.1	9.1	0.9

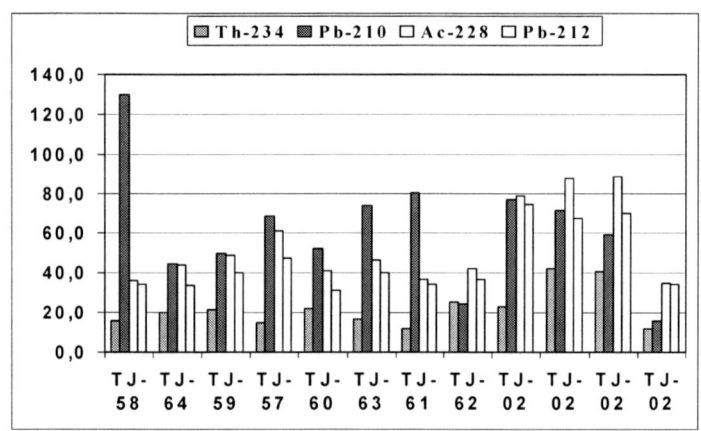
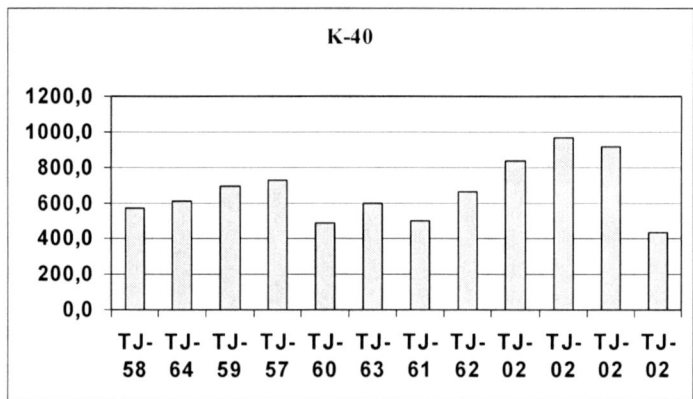
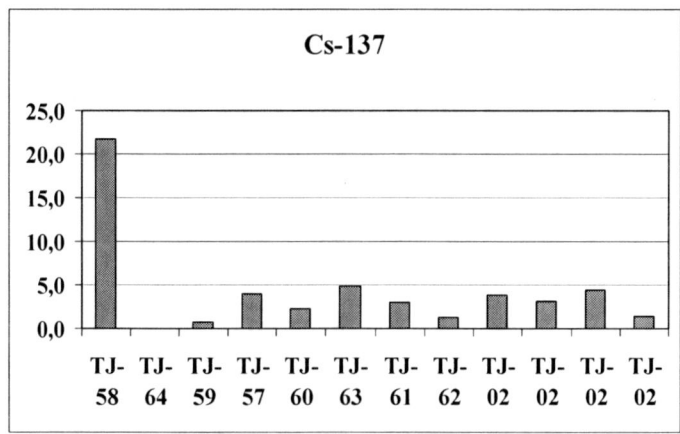

Figure 6. Distribution of natural and technogenic radioactivity in soil samples from foothills of Hissar valley, from west on east

DISTRIBUTION OF NATURAL AND TECHNOGENIC RADIOACTIVITY 163

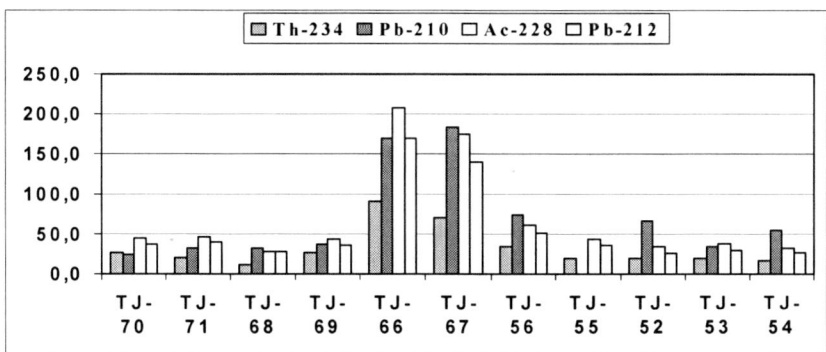

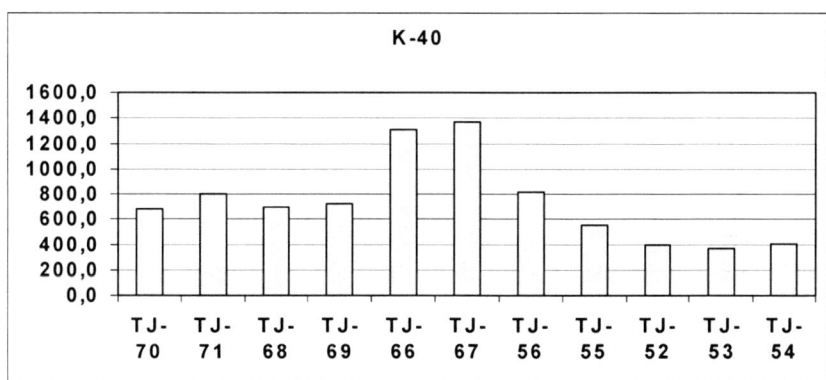

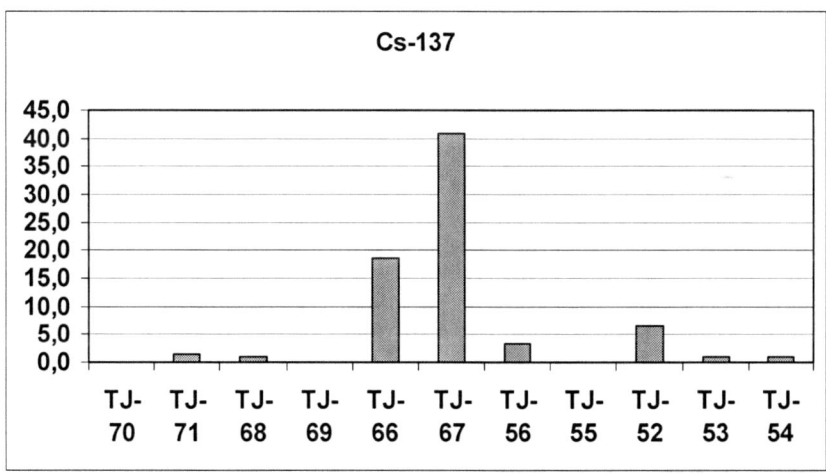

Figure 7. Distribution of natural and technogenic radioactivity in soil samples from the Hissar range, from west on east

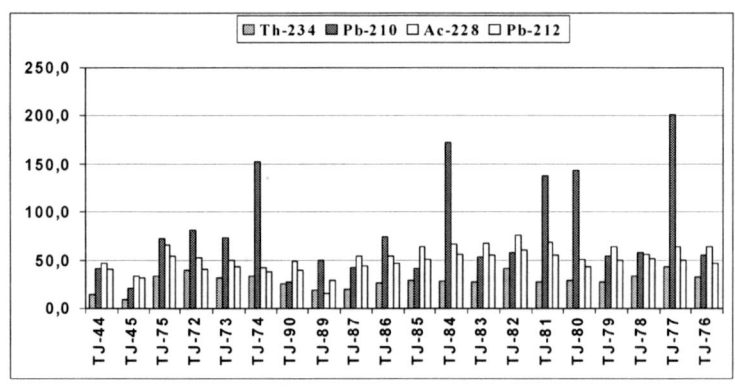
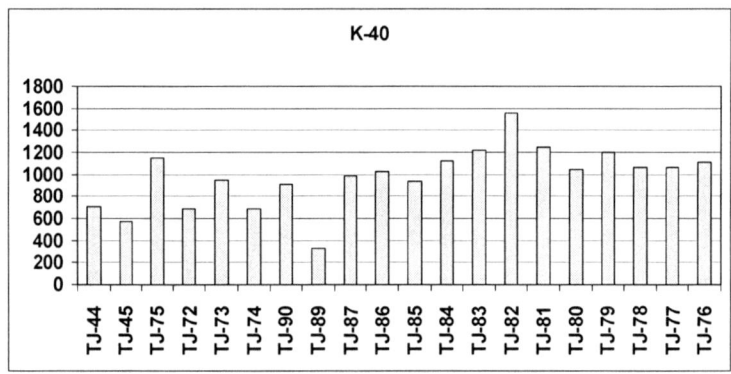
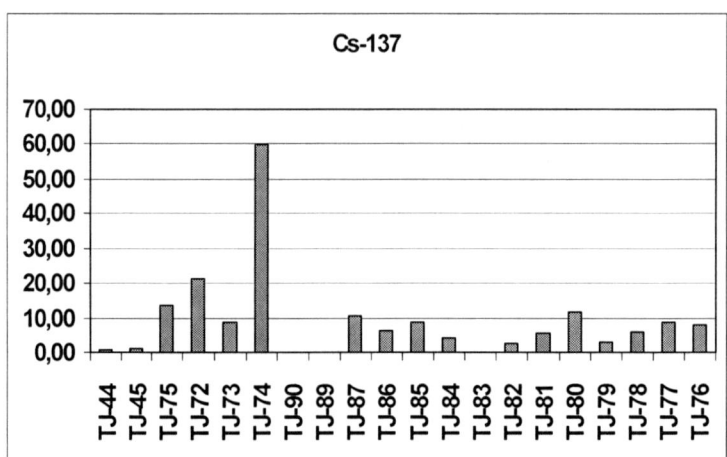

Figure 8. Distribution of natural and technogenic radioactivity in soil samples from the Turkestan and Zarafshan ranges, from west on east

DISTRIBUTION OF NATURAL AND TECHNOGENIC RADIOACTIVITY

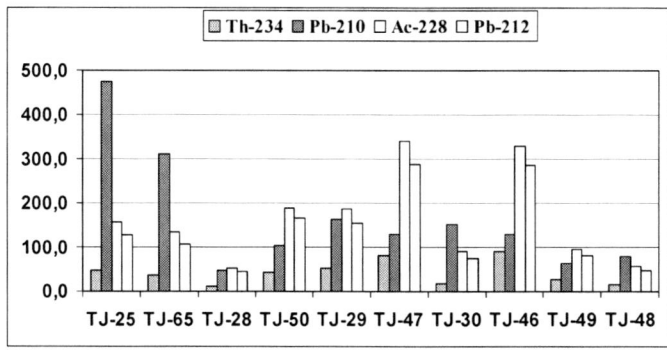

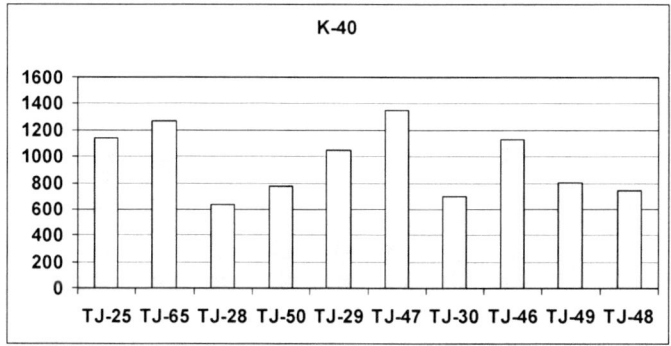

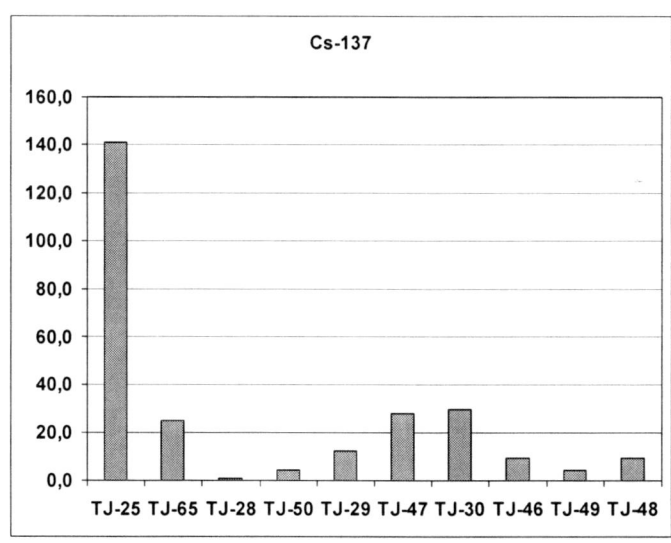

Figure 9. Distribution of natural and technogenic radioactivity in structure in soil samples from the gorge of the Varzob River, from north on the south

CHAPTER 15

QUANTITATIVE ASSESSMENT OF THE MAN-INDUCED URANIUM IN THE TAIL DISPOSAL OF KARA-BALTA MINING PLANT

V.M. ALEKHINA, I.A. VASILIEV, S. IDRISOVA,
S. MAMATIBRAIMOV, G.M. TOLSTIKHIN
Institute of Physics NASc KR
265a, Chui prospect, Bishkek, 720071
Kyrgyz Republic

Abstract: The Hydro-Metallurgical Plant (HMP) of Kara-Balta Mining Plant, Kyrgyzstan, was created in 1950 as a complex of mining and processing enterprises of the atomic branch of the former Soviet Union. A range of different metal processing has taken place at this site over the years, giving rise to large outputs of solid and liquid mineral waste, socalled "tailings". Liquid tailings, consisting of sand, silt, process solutions and water used for transport of this material through the pipeline, are the main waste materials of the Hydrometallurgical plant production (HMP). In the present work, uranium, thorium and radium concentrations together with Ra/U element radios have been determined in the tailing-ponds. These results have been used to calculate the percentage of extracted uranium. The results show a high percent of uranium extraction (up to 97%), and at minimum 92.4%, which is mostly due to the extraction processes utilized in the years 1971 to 1975.

Keywords: uranium, thorium, radium, uranium tailings, Kara-Balta Mining Plant, Kyrgyzstan

1. Background and History

The Hydro-Metallurgical Plant (HMP) of Kara-Balta Mining Plant in Kyrgyzstan was created in 1950 as a complex of mining and processing enterprises of the atomic branch of the former Soviet Union. It functioned on the basis of mining at the Kadji-Sai and Kavak uranium deposits. By 1955, after the end of the construction of the Hydro-Metallurgical plant, Heat Power Plant and auxiliary processes in Kara-Balta town, the HMP became a large enterprise producing uranium concentrate.

The plant also started the production of co-extracts from the molybdenum ores; ammonia paramolybdate. Alongside this, the plant processed molybdenum-containing wastes. All the molybdenum production was ensured at the expense of the wastes and the processing of industrial products, supplied by over 20 enterprises in the former USSR. Together with traditionally produced ammonia the technology for producing other products were also initiated, e.g., the production of rhenium extracts from process products of other industries.

In 1989 the uranium ores was shut down in Kyrgyzstan. The processing of the tin-tungsten concentrate and gold-containing ores and concentrates became the main activity of the mining plant. In recent years the processing of the Kumtor gold was also carried out at the plant. Subdivisions of auxiliary production used to produce the compositions of rare and radioactive metals were also initiated, e.g., plants for drifting rise entries, chisel machine tools, the resin-technical products, ferroconcrete and wood-work products, respirators, electric motors, the anti-friction molybdenum composition "Molycom" for motor and transmission oils concentration. Main suppliers of uranium ores or half-finished products in the 1950s were "Vosmerka" and Kavak. Now, Kazakhstan is the main supplier. With the disintegration of USSR, the uranium production has decreased at the Kara-Balta Mining Plant.

Liquid tailings, consisting of sand, silt, process solutions, and water used for transport of this material through the pipeline, are the main waste materials of the Hydrometallurgical plant production (HMP). The HMP was specialized in processing of uranium ores extracted from the deposits in Kyrgyzstan, Kazakhstan and Russia. Maximum productivity of the plant was around 1.5 million tonnes of uranium-containing ore.

In the 1970s, due to closing of the plants in Kirgizstan and the building of plants in Kazakhstan and Russia, the uranium mining in the South Kazakhstan has changed. They are implementing a new and advanced method, using underground leaching for obtaining concentrates with 30–50% uranium. The uranium supply was stopped at HMP, and in 2005 no uranium-containing effluents were dumped in the tail disposal at these sites. However, new production will probably be imitated.

2. Description of Sites

The tail disposal of KBMC has been exploited since 1954. It is located 1.5 km from Kara-Balta town and 600 m to the south of HMP. Total area of the tail disposal is around 240 ha. It is alluvial, flat and consists of 10 ponds of different size. The old spent ponds are buried beneath the pond No 7. Only pond No 8, having insufficient anti-filtration cover, is functioning now. In all ponds the floor is made of 1 m thick anti-filtration cover of the rammed clay material. Only pond No 9 and its basis pond No 10 (for collection of cyanide-containing effluents of gold-processing and affinage production) are covered with the anti-filtration film. The walls are lined with spent ores of the ore-dressing plant. In northern part of a Sanitary-Protective Zone (SPZ) of the tail disposal site, a dam and a ditch were built to avoid the tail disposal break. The dam is 3–12 m in height, the width on the crest is 5 m, and the ditch depth is 10 m. The dam and the ditch are capable to accumulate "tailings" in a volume of 3.5 million cubic meters. In the tail disposal body a three-layered horizontal drainage is constructed to collect infiltration water.

The clarified liquid phase of the waste pulp was supplied through the drainage system and water-intake facilities of the tail disposal site to the concrete sectional precipitation tank of 10,000 m^3 with accumulating capacities of 130,000 m^3.

3. Metal Processing and Waste Disposal

The Hydro-metallurgical plant (HMP) is not only the plant processing of ores and of the uranium concentrates underground leaching, but also a plant for tin-tungsten-molybdenum concentrates and section for waste processing associated with gold production. The processed ores by the plant comprise: SiO_2 (56–74%), Al_2O_3 (11–14.7%), Fe_2O_3 (1.3–4.6%), CaO (4%, 3%), MgO (2%, 7%), MnO (2%, 3%) etc. The ores also contain solid (short and long-lived) and gaseous radioactive elements. When processing the ores, the following reagents were used: nitric and sulfuric acids, pyrolusite, ammonia, lime, etc. The volume of the above mentioned reagents, especially for a period of 1970–1980, was sufficiently high.

About 6.3 millon cubic meters per year of effluents were dumped in the tail disposal, and the infiltration amounted to about 4 million m^3/year. So, for 1972 about 4 million cubic meters of industrial effluents with mineralization up to 10 g/l came into the aquifer.

The sequence of the ponds filling was as follows. In 1959–1965 the integrated ponds No 1–7 having territory of 65.3 ha were filled. From 1971 to 1975 the silt ponds No 1 and No 2 were filled, having an area of 12.24 ha and 23.45 ha, respectively. According to one of the oldest employees of the plant, when building these ponds, no shielding even of clays was made. Since 1968 until present, the pond No 8 with an area of 57.25 ha has been filled. The pond floor was made of 1 m thick clay material. This pond is not full, but only filled to about 62% of its maximum volume. For the period 1985–1991 the filling of the pond No 9 was carried out. The pond floor was made of the rammed clay material and a film cover. This pond does not operate today because it has been handed over to the joint-stock-company "Kyrgyzaltyn", which is not using it.

During the total time of operation, about 170 million cubic meters of effluents were dumped into the tail disposal sites. Daily dumping of the liquid tailing varied from 11,000 to 20,180 m^3 during the period of maximum ore processing, 1970–1982.

According to data from KBMC by 2005, about 40 million cubic meters of "tailings" of uranium-containing ores were stored at the tail disposal. Design volume for storing makes 62 million cubic meters. When producing 1 t of uranium from uranium-containing ores, 6,000 t of waste were washed in the tail disposal. The solid part of which amounted to 800–1,000 t, contained radioactive uranium daughter-products such as radium-226, radium-222, coming with the supplied ore. Now, when producing 1 t of uranium from the uranium concentrates by the advanced technique, 20–25 t of wastes are washed into the pond No 8, from which a solid part makes 4–5 t. Thus, processing the uranium concentrates has reduced the waste output 250 times.

Solid waste, which was formed at the plants (sand –80% and sludge –20%) during the uranium extraction from raw mineral materials, contained about 70% of undissolved radioactive isotopes. Together with liquid waste, they were dumped into the tail disposal sites. Therefore, the concentration of radioactive isotopes in solid waste (tailings) depends on the scheme of the uranium leaching technique.

4. Results

Investigation of radium and thorium distribution, by stages of the technological cycle, has shown that at the plants with sulfuric acid leaching, about 0.25–0.7% of radium and 80% of thorium in the initial ore are dissolved [1]. When using sodium leaching, the thorium is practically not dissolved, while dissolved radium reaches 1.5–3.0%. Summing up the results, the main part (97.0–98.5%) of radium remains undissolved even when using sodium leaching.

It is also necessary to remember that the tail disposals and dumps also consist of uranium- and thorium daughter products, which are the main sources of environmental radioactive contamination and also gives rise to ^{222}Rn emission.

So, wastes from the Hydro-Metallurgical plant, concentrated in the tail disposal sites, contain natural radioactive isotopes incorporated in the initial uranium mineral materials. Only the uranium isotopes get into the end product, and are therefore present in the waste, at lower concentration than the other natural radioactive nuclides.

In order to determine the quantity of uranium disposed into the "tailings", three boreholes were drilled at the silt pond in the framework of the ISTC project KR-715. It was assumed that 98% of the radium was undissolved and disposed into the "tailings". It was also assumed that radioactive equilibrium existed between ^{238}U and ^{226}Ra in the initial raw materials. Therefore, it was possible to determine the quantity of the remaining technological extracted uranium. Table I presents the results for the concentrations (Bq/kg) of the main radioactive elements, the measured Ra/U ratio (in balanced units), and the amount of technological extracted uranium.

The results show a high percent of uranium extraction (up to 97%), and at minimum 92.4%. This is mostly due to the extraction processes used in the years 1971 to 1975. It is necessary to consider that the minimum value of extracted uranium is observed in the surface layer. This can be explained by some soil drifts from 1975 until now. When necessary, this method can be used to determine the uranium losses as well as the extraction effectiveness of other uranium-processing plants in Kyrgyzstan, where the tail disposals have a negative water balance, i.e. where the radionuclide migration is restricted.

TABLE I. Concentrations of radionuclides in core samples

№	Sample code	Sampling depth, m	U Bq/kg	^{232}Th, Bq/kg	Ra Bq/kg	Ra/U	Extraction U (%)
1	Borehole1-0	0.2	500	40	5,304	10.6	92.4
2	Borehole 1-1	1.5	500	32	7,600*	15.2	95.3
3	Borehole 1-2	3	538	25	9,898	18.4	96.5
4	Borehole1-3	4.5	612	49	8,143	13.3	94.4
5	Borehole 1-4	6	612	44	9,057	14.8	95.1
6	Borehole 1-5	7.5	612	40	11,953	19.5	98.8
7	Borehole 1-6	9	625	60	13,556	21.7	97.3
8	Borehole 1-7	10.5	560	49	11,430	20.4	97.0
9	Borehole 1-8	12	612	52	15,473	25.3	98.0
10	Borehole 1-9	13.5	612	52	14,208	23.2	97.6
11	Borehole 1-10	15	640	49	15,360*	24.0	97.8
12	Borehole 1-11	16.5	688	73	16,513	24.0	97.8
13	Borehole 1-12	18	612	48	12,376*	20.2	97.0
14	Borehole 1-13	19.5	625	21	8,238	13.2	94.3
15	Borehole 1-14	21	675	17	12,603	18.7	96.6
16	Borehole 1-15	22.5	740	24	15,976*	21.6	97.3
17	Borehole 1-16	24	688	24	19,350	28.1	98.4
18	Borehole1-17	25.5	–	–	1,964	–	–
19	Borehole 1-18	27	–	29	1,254	–	–
20	Borehole 2-0	0.2	538	35	6,176	11.5	93.2
21	Borehole 2-1	1.5	512	48	11,144	21.8	97.3
22	Borehole 2-2	3	538	48	16,652	31.0	98.7
23	Borehole 2-3	4.5	600	52	17,700	29.5	98.6
24	Borehole 2-4	6	612	57	15,636	25.5	98.0
25	Borehole 2-5	7.5	612	48	20,210	33.0	98.8
26	Borehole 2-6	9	525	45	17,002	32.4	98.9
27	Borehole 2-7	10.5	600	48	13,852	23.1	97.6
28	Borehole 2-8	12	612	52	21,878	35.7	97.2
29	Borehole 2-9	13.5	612	57	11,576	18.9	96.6
30	Borehole 2-10	15	600	31	9,316	15.5	95.5
31	Borehole 2-11	16.5	525	21	8,389	16.0	95.6
32	Borehole 2-12	18	640	27	9,245	14.4	95.0

(*Continued*)

TABLE I. *(Cont.)*

33	Borehole 2-13	19.5	538	24	8,509	15.8	95.6
34	Borehole 2-14	21	600	18	6,342	10.6	92.4
35	Borehole 3-0	0.2	488	29	6,979	14.3	94.9
36	Borehole 3-1	1.5	512	37	10,972	21.4	97.3
37	Borehole 3-2	3	500	40	11,526	23.0	97.6
38	Borehole 3-3	4.5	625	48	18,331	29.3	98.6
39	Borehole 3-4	6	525	50	19,007	36.2	99.2
40	Borehole 3-5	7.5	550	40	18,071	32.8	98.9
41	Borehole 3-6	9	575	51	22,649	39.4	99.4
42	Borehole 3-7	10.5	625	48	29,555	63.2	99.9
43	Borehole 3-8	12	538	29	19,958	37.1	99.3
44	Borehole 3-9	13.5	575	44	13,747	23.9	97.8
45	Borehole 3-10	15	612	29	13,355	21.8	97.4
46	Borehole 3-11	16.5	612	19	12,257	20.0	96.9
47	Borehole 3-12	18	800	14	14,608	18.3	96.4
48	Borehole 3-13	19.5	625	–	15,228	24.4	97.8
49	Borehole 3-14	21	–	–	1,219	–	–
50	Borehole 3-15	22	–	–	483	–	–
		Minimum:	488	14	483	10.6	92.4
		Mean:	593	40	13,096	23.3	97.0
		Maximum:	800	73	29,555	63.2	99.9

* – interpolation data

5. Conclusion

By determining the uranium, thorium and radium concentrations and the Ra/U radios in the tailing ponds, the percentage of extracted uranium can be calculated. The results show a high percent of uranium extraction (up to 97%), at minimum 92.4%, which is mostly due to the efficient extraction processes used during the years 1971 to 1975.

This method can be used to determine the uranium losses as well as the extraction effectiveness in other uranium-processing plants in Kyrgyzstan, where the tail disposals are restricted with respect to radionuclide migration

References

1. Smirnov, Yu.V., Yefimova, Z.I., Skorovarov, D.I. et al. (1975) Removing the wastes from the plants on the uranium processing, *Atomnaya tekhnika za rubezhom* 11, 11. (in Russian)

CHAPTER 16

STUDY OF THE ^{222}RN DISTRIBUTION IN AIR ABOVE THE TAIL DISPOSAL SITE AT KBMP (KYRGYZ REPUBLIC) AND ITS POSSIBLE TRANSFER TO ADJACENT TERRITORIES

I.A. VASILIEV, V.M. ALEKHINA,
S. MAMATIBRAIMOV, O.I. STARODUMOV,
S. IDRISOVA, D.A. IVANENKO
*Institute of Physics NASc KR 65a, Chui prospect,
Bishkek, 720071, Kyrgyz Republic*

Abstract: In the present paper the characteristics of an instrument developed for the determination of radon levels in air are described together with results of the ^{222}Rn distribution in the surface layer of the atmosphere above the territory of the Kyrgyz Mining plant (KBMP) tail disposal site. Air transport from the boundary of the KBMP is estimated. For the observation period the radon concentration changed from 7 Bq/m^3 to more than 800 Bq/m^3, with average value of about 100 Bq/m^3. Results also show that radon can be transported hundreds of kilometers, up to Bishkek city and further on.

Keywords: radon distributions, air, tail disposal site, Kyrgyz Republic

1. Radon Measurements, Description of Instrument

The basic instrument used is based on the measurement of scintillation counts when the alpha particles hit the walls of an emanation camera, where the production instruments are the emanometer "Radon" and the alpha-analyzer "Alpha-1". The developed instrument differs from the production instruments by a pulse former and a measuring process excluding the gas delay in the emanation camera, lasting 2–3 h for establishing radioactive equilibrium between radon and its short-lived disintegration products. The pulse former (discriminator) will only let the pulses pass, which on their drop correspond to the time of the scintillator lighting; typical for scintillations caused by the alpha particle. Such an approach reduces the instrument background with a factor of about ten. In the selected air samples, radioactive equilibrium has been established between radon and its short-lived disintegration products. Therefore, it is not necessary to delay the samples in emanation cameras for 2–3 h to establish radioactive equilibrium.

The instrument has two operating modes: "calibration" and "operation". The calibration is needed to check the instrument with standard radioactive materials. The calibration results are presented in decay per minute. The "operation" mode consists of two sub-modes: the expression related to the radon determination in air for 10–15 min, and the monitoring sub-mode allowing the determination of the radon level in air samples automatically on an hourly basis and register the output in the instrument memory. The instrument memory will register the output every hour during an operation time of 1.5 months. Results are given as Bq/m^3. The detection limit of the instrument is about 20 Bq/m^3. The admissible concentration for the radon in dwellings and offices air is limited to 200 Bq/m^3.

2. Results and Discussion

In Table I the results of the radon level determination at a height of one meter above the tail disposal surface at the Kyrgyz Mining plant (KBMP) are presented. The samples

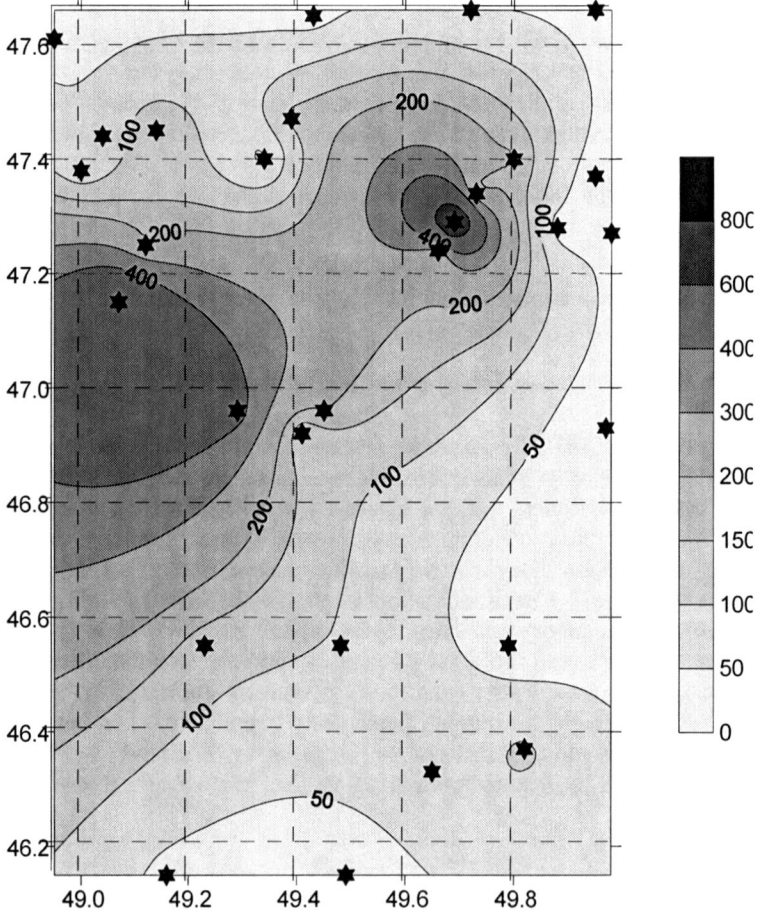

Figure 1. The ^{222}Rn distribution in air at the KBMP tail disposal site. The coordinates (X-and Y-axes) are given in conventional units

TABLE I. Radon level in air at the tail disposal area, the Kyrgyz Mining plant (KBMP)

Order #	The location of the sampling site	Conventional coordinates of the site	Altitude above sea level	Dose rate µR/h	Radon level, Bq/m^3
1	200 m northward of the lodge, metal scrap warehouse	47.27; 49.98	819	50	≤100
2	Borehole 120	47.66; 49.95	804	30	≤55
3	Borehole 119	47.66; 49.72	800	30	≤100
4	Borehole 118	47.65; 49.43	803	60	≤70
5	Borehole 116	47.61; 48.95	804	35	≤70
6	N.W. of the drainage	47.44; 49.04	813	1,090	68
7	60 m to the north of drainage	47.45; 49.14	817	3,040	138
8	300 m eastward of # 7	47.47; 49.39	815	3,040	162
9	N.W. of the silt pond	47.38; 49.00	824	1,030	61
10	Section-line with borehole 118	47.40; 49.34	830	370	34
11	200 m from # 10	47.40; 49.80	834	360	208
12	Borehole 1, drilled	47.34; 49.73		342	202
13	Borehole 2, drilled	47.29; 49.69	844	1,090	872
14	Borehole 3, drilled	47.24; 49.66	838	1,090	248
15	The eastern corner of pond 7	47.28; 49.88	844	1,030	27
16	Turn to pond 8, pond 8	46.93; 49.97	833	340	14
17	Middle of the dam of pond 8	46.96; 49.45	847	370	204
18	Pond 8	46.92; 49.41	851	1100	107
19	Corner of ponds 8 and 9	46.96; 49.29	850	240	445
20	Section of pond 8	46.55; 49.23	856	1,150	127
21	Pipe	46.55; 49.48	855	1,100	96
22	Eastern corner of ponds 8-9	46.55; 49.79	855	1,180	21
23	Eastern corner of pods 9-10	46.37; 49.82	855	1,100	107
24	In the center of ponds 9-10	46.33; 49.65	856	200	85
25	Southern dam, center, top	46.15; 49.49	864	25	19
26	S.W. corner of the tailing dump	46.15; 49.16	867	400	38
27	South of the silt pond	47.15; 49.07	824	2,300	607
28	North of the silt pond	47.25; 49.12	823	1,100	217
29	Ash pond	47.37; 49.95	866	18	≤35

were taken October 2, 2003. The radon level was determined by the emanation method on the installation "Alfa-1" [1, 2] and with the above described instrument developed in Radiometric Laboratory of IPh NASc KR. [3, 4]. The radon distribution in air, 1 m height above the tail disposal surface, is shown in Fig. 1. Figure 2 presents the observed dose rate of gamma radiation (μR/h) and radon level (Bq/m^3). A clear correlation between the γ-radiation dose rate and the radon concentration in air was not obtained. This is explained by different emanation ability to cover the ponds of the tail disposal. Following the data in Table I and Figure 1, increased radon levels were observed at the "Silt" pond and at the northern part of the pond 7.

The radon levels at the "Silt" pond of the tail disposal are given in Table II. Near ground (0.05 m), the radon concentrations reached up to 970 Bq/m^3; at a height of 0.5 m above surface, it decreased more than two times and then it did not change up to the height of 10 m (80% of the values at a height of 0.5 m). The measurements were carried out under calm wind conditions. Therefore, under windy weather conditions the radon could be transported over long distances. For studying the air transport process at the Kara-Balta meteorological station, an instrument for automatic determination of the radon level I air was installed [3, 4].

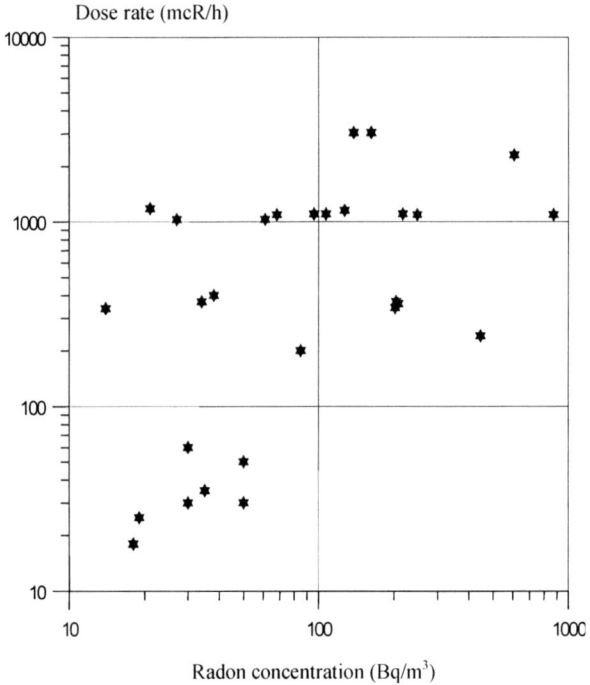

Figure 2. The γ-radiation dose rate and the radon concentration

TABLE II. Radon level in air at the "Silt pond" of the tail disposal (10.11.2004)

Sampling height, m	Conventional coordinates	Altitude above sea level, m	Dose rate, μR/h	Radon concentration	
				Bq/m^3	In MAC
0.05	47.38; 49.13	822	1,300	970	4.8
0.20				720	3.6
0.50				380	1.9
1.00				376	1.9
2.00				360	1.8
10.0	47.40; 49.17	832	390	296	1.5

At regular intervals of 1 h, at a height of 1 m above the soil surface level, the air sampling and radon determination were carried out. The measurements were initiated on February 8, 2005, and are still being conducted. As an example, the variation of the radon level from April 4 to April 30, 2005 is presented in Fig. 3.

For the observation period (from the beginning of the measurements up to the preparation of the present article), the radon concentration changed from 7 Bq/m^3 to more than 800 Bq/m^3 (the MAC is 200 Bq/m^3), with an average value of about 100 Bq/m^3. This is higher than the mean concentrations associated with the radon air transport from the KBMP tail disposal. Therefore, we examined the radon concentration dependency of the wind direction, as demonstrated in Fig. 4. Here, the vector length relates to the radon concentration (distance from the centre to the point), and the angle relates to the wind direction (0 – North, 90 – West, 180 – South, 270 – East).

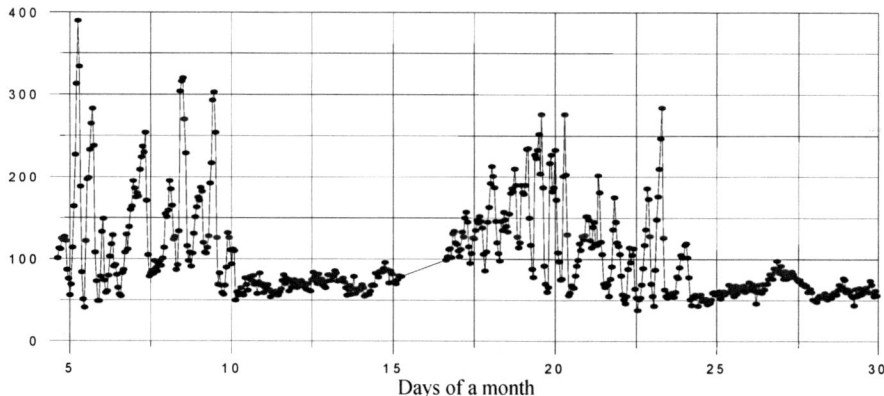

Figure 3. Radon concentration in air at the Kara Balta meteorological station Rn (Bq/m^3)

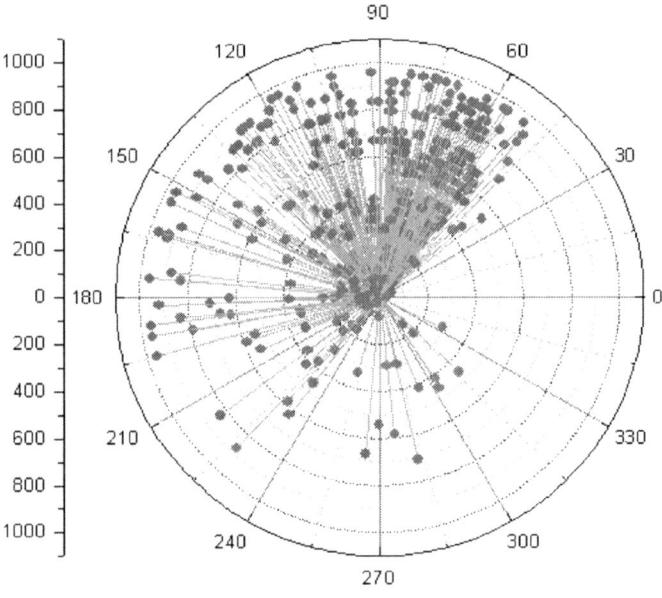

Figure 4. Dependency of the radon concentration in air on the wind direction at the Kara-Balta meteorological station

As seen from the diagram (Fig. 4), the predominant wind direction is from the west to the east. Taking into consideration that the radon half-life is 3.8 days, the predominant wind direction from the west to the east and the relief of the area (narrowing of the Chui valley from west to east), radon can be transported (until it completely decays) for hundreds of kilometers, up to Bishkek city and further on.

It will be possible to obtain a complete picture of the radon air distribution if we install similar instruments all over the territory of the Chui valley.

3. Conclusion

When estimating the ecological situation on the Kara-Balta area in general, on the basis of the adopted sanitary norms, it is possible to state that radon concentration in air, along the boundaries of the Chui valley from the east to the west of the tailing dump, may represent an ecological hazard. Under windy weather conditions, the radon transport may take place over a long distance. It is possible to obtain a complete picture of the radon distribution if similar devices are installed the all over the territory of the Chui valley. For the observation period the radon concentration changed from 7 Bq/m^3 to more than 800 Bq/m^3, with an average value of about 100 Bq/m^3. Taking into consideration that the radon half-life is 3.8 days, the predominant wind direction from the west to the east and the relief of the area, radon can be transported for hundreds of kilometers, up to Bishkek city and further on.

References

1. Shishkin, V.L. (1961) Methods of natural radioactive elements analysis. *Gosatomisdat*, p. 150. (in Russian)
2. Yakubovich, A.L., Zaicev, E.I., Pzhiyalgovskiy, S.M. (1973) Nuclear-physical methods of mineral raw material analysis. *Atomisdat*, p. 392. (in Russian)
3. Vasiliev, I.A., Alekhina, V.M., Idrisova, S., Mamatibraimov, S. (2006) Radioecological and adjacent problems of the uranium production 3, Proceeding, edited by Vasiliev, I.A., Bishkek, Ilim. (in Russian)
4. Vasiliev I.A. (2004) The instrument for determination of radon concentration in air. International Seminar *"Commercialization of ISTC Project results"*. September 13–15, 2004 Issyk-Kul Lake, Kyrgyzstan.

CHAPTER 17

STUDY OF TRANSBORDER CONTAMINATION OF THE SYR-DARYA AND AMU DARYA RIVERS AND THEIR INFLOWS

B.S. YULDASHEV[1], H.D. PASSELL[2],
U.S. SALIKHBAEV[1], R.I. RADYUK[1], A.A. KIST[1],
S.V. ARTEMOV[1], G.A. RADYUK[1], E.A. ZAPAROV[1,]
E.A. DANILOVA[1], A.A. ZURAVLEV[1],
V.S. VASILEVA[1], E.E. LESPUKH[1]
[1] *Institute of Nuclear Physics AS RU*
[2] *Geosciences and Environment Center Sandia National Laboratories, Cooperation, Monitoring Center, USA*

Abstract: Monitoring of an environment in river basins of Syr-Darya, Amu Darya, and their inflows started in 2002 according to the international agreement of scientists of Central Asia, Kazakhstan and the USA. Great volume of experimental data on salt content of water, total alpha-beta-activity concentrations, and radionuclide composition of water samples, bottom sediments, and coastal soils was obtained. At the same time more than 28 elements were determined in these samples by using Instrumental Neutron Activation Analysis (INAA). Data analysis showed that the distribution of total alpha-beta-activity correlates with the distribution of radionuclides, but they show that there are additional technogenic sources of heavy radionuclides.

Keywords: radionuclides, metals, Syr-Darya River, Amu Darya River, water, sediments, soils

1. Introduction

Radiation monitoring of the environment in the river basins of Syr-Darya, Amu Darya started in 2002 in compliance with international agreement of the scientists of Central Asia, Kazakhstan, and USA [1]. The first results from the work on Syr-Darya river monitoring have been presented earlier [2], showing data on salt content and alpha-beta-activity concentrations of water, bottom sediments, and coastal soil samples. It was shown that the salt content and the alpha- and beta-activity concentrations in water have regular and same character depending on the remoteness from the riverhead. Alpha- and beta-activity concentrations in the bottom sediments and coastal soils have also the same and regular character, however, absolutely different from what was observed in the water samples. Then, data on natural radionuclides in samples collected along the river were analyzed [3]. It turned out, that the distribution of thorium and uranium in soil and bottom sediment samples correlate, as a whole, with the behaviour of alpha-and beta-activity concentrations, but have distinctive features associated with anthropogenic activity of man. As a whole, it became clear that it was necessary to continue the monitoring and increase the number of sampling locations. In 2003, a 3-year project on radioactive monitoring of transborder rivers of Central Asia and Kazakhstan, Syr-Darya, Amu Darya, and their inflows started. During the first half of the project the information on the radionuclide composition in soils and bottom samples along the rivers of Uzbekistan was obtained. By averaging activities of the nuclides by seasons we got average seasonal distribution of radioactivity in all big rivers of Uzbekistan and found the areas with elevated concentrations [4]. Example of such distribution is shown in Fig. 1. In the area of elevated radionuclide levels we observed possible seasonal changes in the activity concentrations. As a result of this, after five seasons we conducted seasonal analysis of the radionuclide distribution in the areas of elevated concentrations.

In this paper we present the results of such analysis and show the changes in the radionuclide activity concentrations depending on the season.

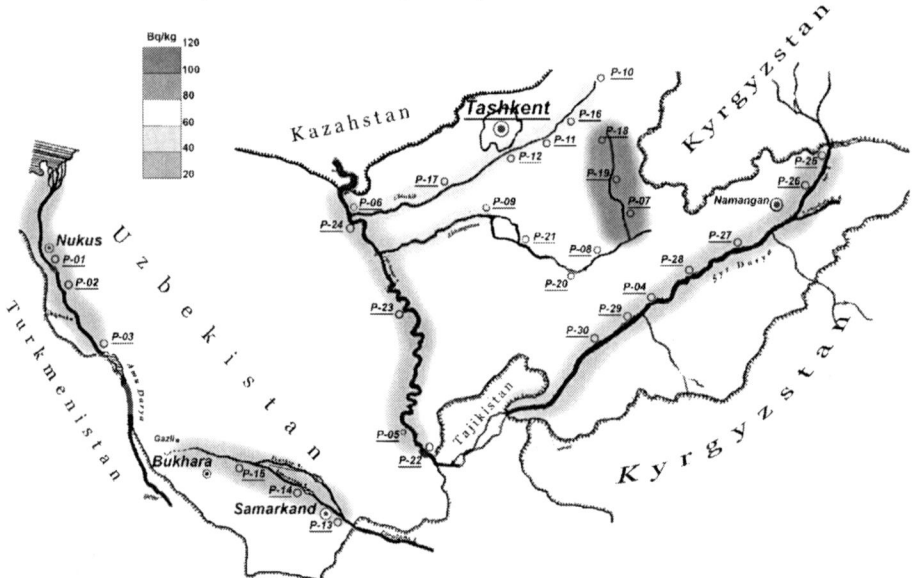

Figure 1. Average activity concentrations of thorium in bottom sediments and coastal soils along the rivers of Uzbekistan

2. Experiment Techniques

Analysis of the radioecological situation was done based on alpha- and beta-activity data and the radionuclide composition. Total alpha- and beta-activities were obtained by using gas-flow counter of SOLO system. Gamma spectres were obtained by using germanium rectifiers. For SOLO standard weight of samples were 10 g. Exposition time for alpha counts were 180 min, for beta counts – 20 min. Gamma spectres were measured by volumetric geometry. Samples of 300–1,400 g were transferred to Marinelli vessels. Depending on the weight of samples, corrections were made on self-absorption by changing the curves of effectiveness. Curves of effectiveness were obtained by cascade nuclides, which are contained in samples with normalization on known activity concentrations of natural KCl salt. Exposition times on gamma detectors were from 6 to 12 h.

We analyzed seasonal changes of activity concentrations in the area of increased concentrations in the Akhangaran River with its inflow – Kuvasay. Sampling locations were 07, 08, 09, 18, 19, 20, 21 and 31 (new location at the confluence of the Akhangaran and Syr-Darya rivers). At the same time we observed the radionuclides K-40, Pb-214 and Ac-228. The last two nuclides represent uranium and thorium radioactive chains, respectively, and in the gamma spectres the gamma-lines from these radionuclides are the most intensive. At each sampling location we collected samples of bottom sediments and coastal soils twice a year (in spring and fall). The analytical results are expressed as activities per unit weight of dry sample, which was preliminarily dried, crumbled up, and sieved (Bq/kg).

3. Results

All measurement results of radionuclides in bottom sediments are presented in Figs. 2 and 3. Standard deviations in the determination of gamma and beta activities are

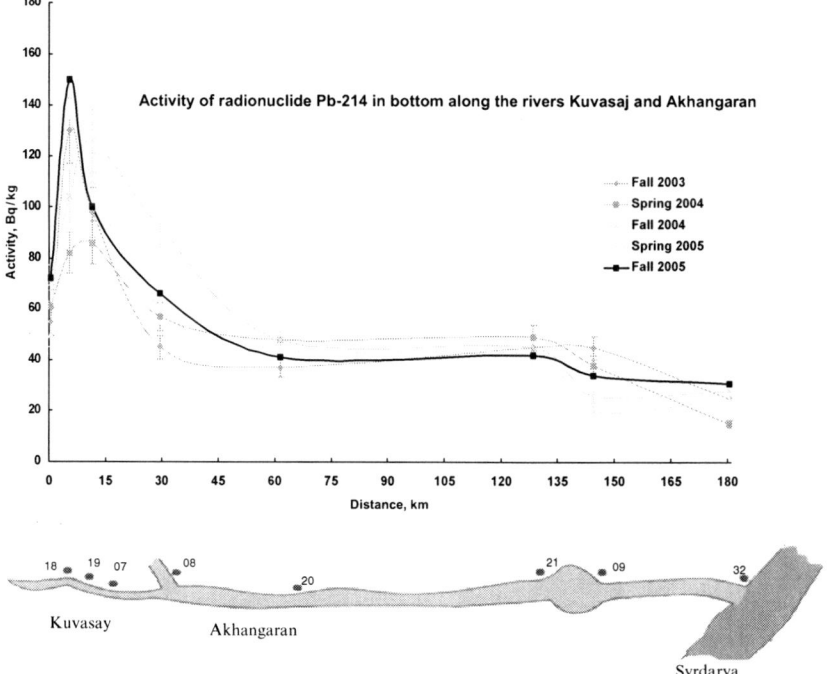

Figure 2. Activity concentration of Pb-214 in bottom sediment samples along the Akhangaran with Kuvasay inflow

10%. In determination of alpha activities standard deviations can reach 20% for low-active samples. Quick analysis of the material shows that the highest activity concentrations were identified in the sampling locations 07, 18 and 19 for all seasons, and in these locations the activities significantly change by seasons. Here, the most significant change was seen for the activity concentrations of the daughter nuclides of uranium and in the total alpha-activity. We will continue to monitor these activity changes.

4. Bottom Sediments

The activity distribution of Pb-214 in bottom sediment samples along the Akhangaran with Kuvasay inflow is shown on Fig. 2. Maximum activities were observed in locations 19 and 07 for all seasons. In fall, maximum activities were observed in location 19, and in spring maximum activities were lower then down stream in location 07. This tells us about the migration of nuclides in the uranium decay chain. Activities in all seasons decreased to 30–50 Bq/kg 60 km downstream and stayed at this level over 67 km until the reservoir (location 21). After that, the activity concentrations went further down to 20–30 Bq/kg at the confluence of the Akhangaran River and Syr-Darya River. By looking at displacement of maximum activities from location 19 till location 07 during 1.5 years, the migration rates of radionuclides in the uranium chain on the river bottom (Kuvasay inflow) have been estimated to about 4 km per year, at this section.

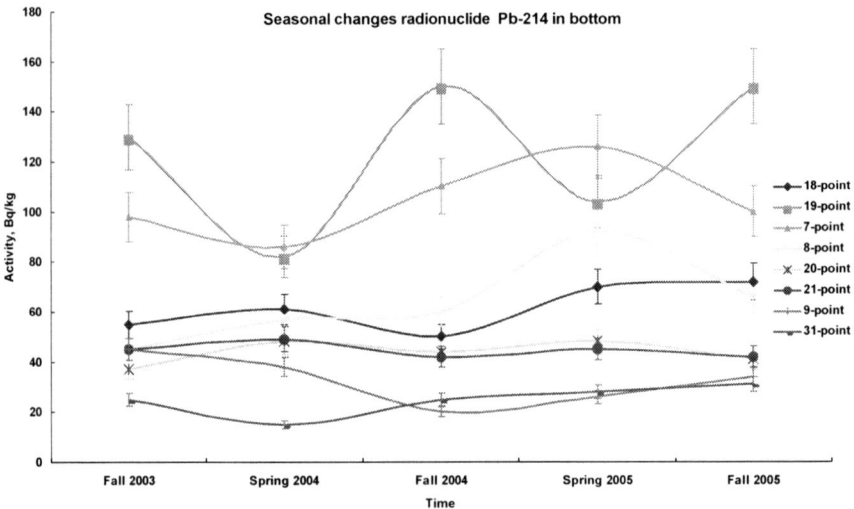

Figure 3. Seasonal change of activity of radionuclide Pb-214 in bottom sediments in the Akhangaran

Figure 3 shows the change in the activity concentrations of Pb-214 in the investigated locations depending with time. The greatest change is in location 19: every fall, the concentration increased compared to spring activity by 1.5 times (and decreased every spring). Changes were also seen in the remaining sampling locations, but not to the same extent: in locations 07 and 08, closest to active location 19, it is a clear changes with time, while in locations 18 and 20 the level stays almost constant during the year. Also the total alpha activity concentrations in the bottom sediments showed evident seasonal dependence. The alpha activity concentrations changed along the river from maximum level of 3,700 Bq/kg in location 07 till 870 Bq/kg at the entry into Syr-Darya. In Fig. 4 the total alpha activities are given.

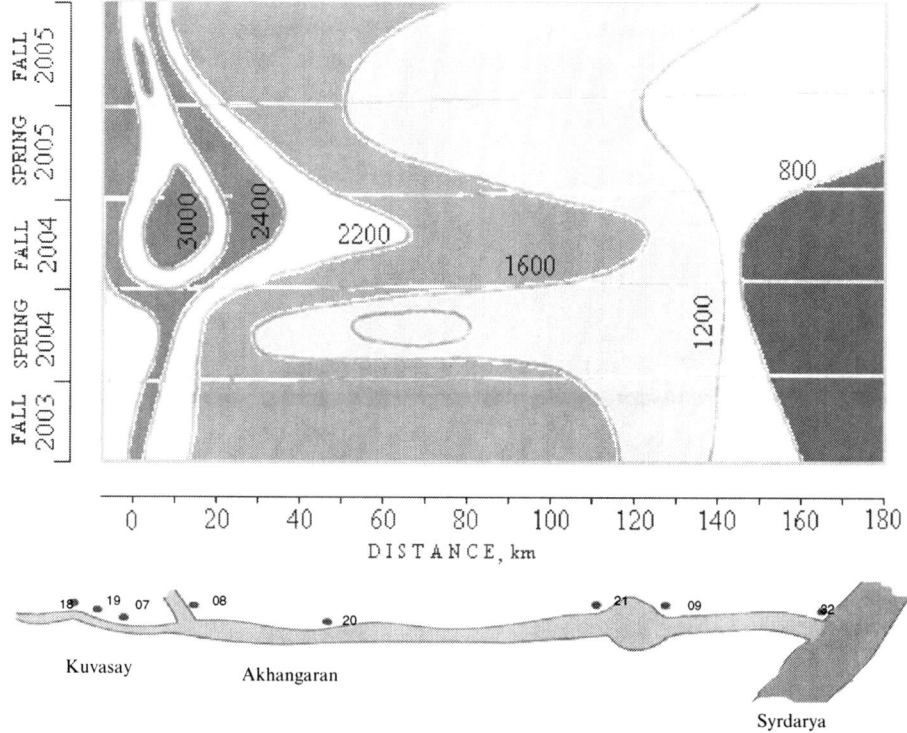

Figure 4. Diagram of the seasonal changes of alpha-activity concentrations of bottom sediments in the Akhangaran River

Here it can be seen that in fall 2003, 2004, and 2005 the levels of activities higher than 1,500 Bq/kg went down the river much further than during springs. Maximum activities appeared also in fall. In fall 2003 the alpha activity concentrations increased till about 2,500 Bq/kg in the area of locations 19 and 07. In fall 2004 in the area of location 07 activity concentrations went up to more than 3,500 Bq/kg. In fall 2005 it reached more than 3,000 Bq/kg. Figure 5 shows the beta activity distribution in bottom sediments in the rivers of Uzbekistan.

5. Coastal Soil

The concentration of Pb-214 in soil samples collected during five seasons is shown in Fig. 5. In general, the levels of radioactivities in soils were similar to the ones of the bottom sediment activities: maximum activities were observed in locations 19 and 07 and the activity decreased from the sampling location 20 to the level of about 20 Bq/kg (same as for bottom sediments). However, there was no exact seasonal correlation. Twice the maximums of the activity were observed in location 19, but once in fall and other times during spring. It is known that organic substance of soil significantly influences accumulation of radionuclides in soil, their sorption by colloidal particles, and ability to exchange elements between soil, soil solution, and plant roots [4].

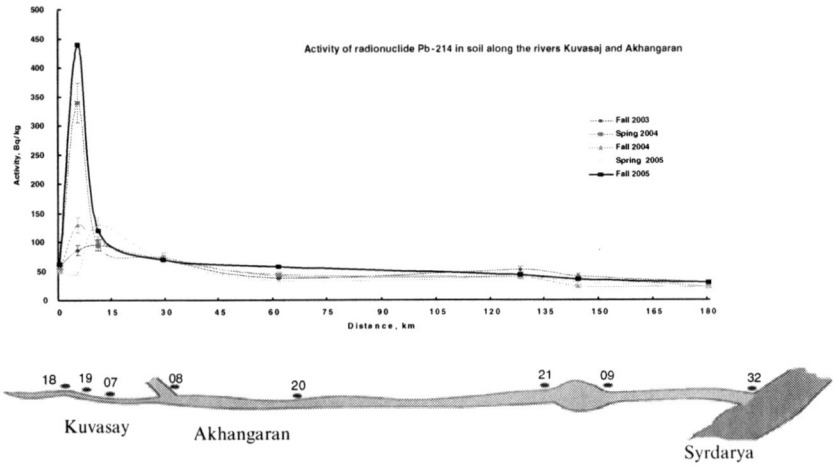

Figure 5. Activity distribution of Pb-214 in soil samples along the Akhangaran River and Kuvasay inflow.

Radionuclides became solute, both as inorganic compound and complex organ metallic compound of humus. Levels of activities of Pb-214 in soil was higher than in bottom sediments: in spring 2004 there was a drastic increase of Pb-214 till 340 Bq/kg, and in fall 2005 it increased till 440 Bq/kg. However, decrease of activity concentrations was observed down the river similarly to bottom sediments.

In Fig. 6 the changes of the Pb-214 activities with time in the sampling locations are shown. Similarly to bottom sediments, the greatest dynamic were observed at sampling location 19. However, the pattern of activity change at location 19 was completely different; there was a drastic activity increase in spring 2004 and in fall 2005.

The activity concentrations reached maximum value in location 07 in spring 2005. If this increase was caused by migration of nuclides from location 19 in spring 2004, then the rate of this migration is 6 km per year. This is slightly higher than the migration of

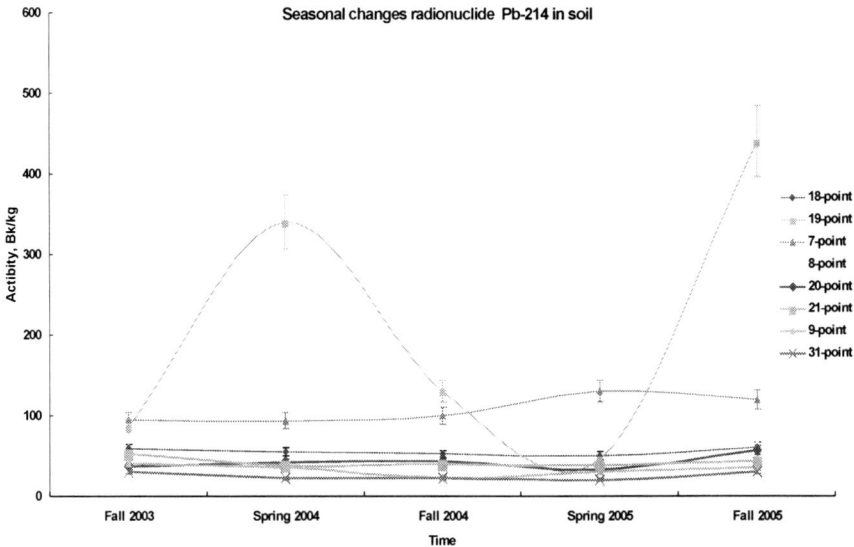

Figure 6. Seasonal changes of activities of Pb-214 in coastal soil of the Akhangaran

nuclides in bottom sediments. Small increase of activity was observed in the rest of the locations. Total alpha-activity in coastal soils (Fig. 7), similarly to bottom sediments, obviously, depended on seasons, however, there is no exact seasonal correlation. Activity concentrations in soil were at its maximum in spring. Large increase in the alpha activity concentrations were observed in spring 2004 (up to 5,700 Bq/kg) and in spring 2005 (up to 5,800 Bq/kg) at sampling location 19. If the second increase in the activity was caused by migration of radionuclides from location 19 along the river, then its rate is 6 km per year. Possibly, the great increase in soil activity at location 19 in 2004 caused drastic increase in the bottom sediment activity in fall 2004, but at location 07. Then the rate of radionuclide migration from the bank and into the river would be the double – 12 km per year. Note again the difference in the distribution of activities in soils and bottom sediments. In bottom sediments significant changes in activity concentrations were observed during the same season as in the riverhead. In soil, significant changes in activity concentrations were found in the next season after high activity was observed in riverhead. Most likely this difference is caused by slower migration rate of radionuclides along the bank than on the river bottom. Based on activation analysis of samples of bottom sediment and soil 28 different elements were determined. For most of the elements, they were found in cross correlation in the bottom sediment – soil system. Reduced number of correlations observed for the spring sampling is, probably, caused by dilution of water and leaching of elements from bottom sediment in spring high water events.

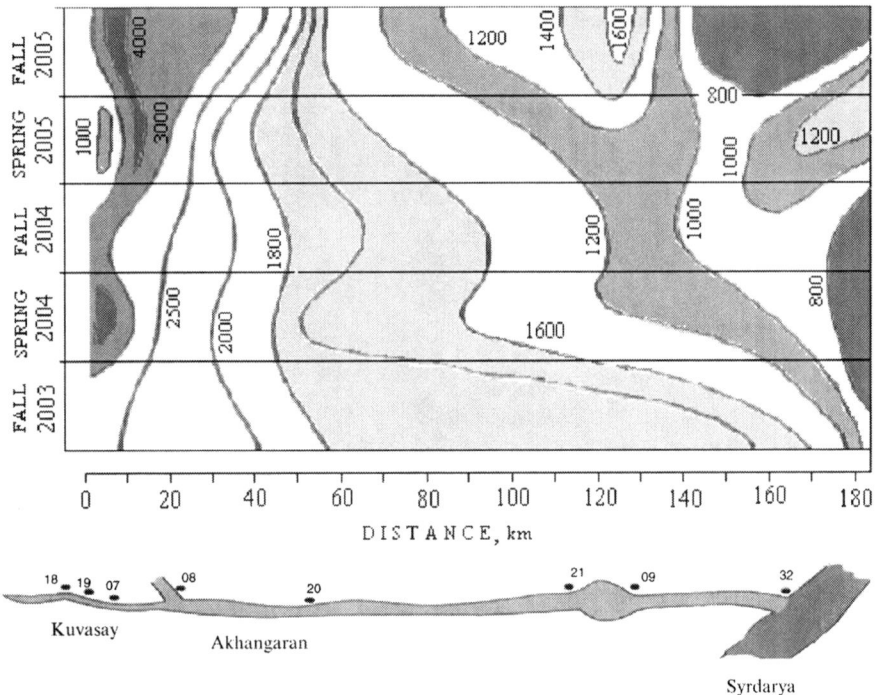

Figure 7. Seasonal changes in the alpha- and beta-activities in coastal soil along the Akhangaran River

6. Conclusion

The result obtained in this work, showed that the concentrations of natural radionuclides in coastal soil and bottom sediments along the rivers of Central Asia conform with their natural levels in soil. The activity concentration ranged from 20 to 40 Bq/kg for the uranium chain radionuclides and 40–370 Bq/kg for potassium [5]. Elevated concentrations of natural radionuclides were observed along the inflow of the Akhangaran River – Kuvasay, where the radionuclide levels along a great section of the river was higher (approx. two times higher) than their levels in Syr-Darya. Significant seasonal variations in the content of natural radionuclides were identified only in one location: the inflow of the Akhangaran River – Kuvasay, and seasonal displacements of radionuclides were observed downstream the Akhangaran River. During 2.5 years of systematic observations no drastic changes were revealed in total amount of radionuclides in the chosen rivers. The main conclusion is that activity concentration in bottom sediment may be considered as averaged information, while water gives the more "at date" information.

References

1. Passell H.D. and et al. (2002) The Navruz Project: Transboundary Monitoring for Radionuclides and Metals in Central Asia Rivers, Sampling and Analysis Plan and Operation Manual. Sandia National Laboratories, p. 85
2. Barber D.S., Yuldashev B.S., Passell H.D., et al. (2003) Radiation Monitoring of Syr-Darya River, Eurasia, Nuclear Bulletin, #2, pp. 82–87
3. Yuldashev B.S., Passell H.D., et al. (2004) Radiation Monitoring of Syr-Darya River (II)
4. 3rd Eurasia Conference Nuclear Science and Its Application. Tashkent Uzbekistan, p. 192
5. Yuldashev B.S., Salikhbaev U.S., Radyuk R.I., Vdovina E.D., Kist A.A., Danilova E.A., Artemov S.V., Radyuk G.A., Zaparov E.A., (2005) Radioecological Monitoring of Syr-Darya and Amu Darya Rivers. International Workshop on Radiological Sciences and Applications, Peaceful Uses of Nuclear Technologies, p. 38

CHAPTER 18

THE NAVRUZ PROJECT: COOPERATIVE, TRANSBOUNDARY MONITORING, DATA SHARING AND MODELING OF WATER RESOURCES IN CENTRAL ASIA

H.D. PASSELL[1], V. SOLODUKHIN[2],
S. KHAZEKHBER[2], V.L. POZNIAK[2],
I.A. VASILIEV[3], V. ALEKHINA[3], A. DJURAEV[4],
U.S. SALIKHBAEV[5], R.I. RADYUK[5], D. SUOZZI[1],
D.S. BARBER[1]
[1] *Geosciences and Environment Center Sandia National Laboratories, Cooperative, Monitoring Center, USA*
[2] *Institute of Nuclear Physics NNC RK, Ibragimov str. 1, 050032, Almaty, Kazakhstan*
[3] *Institute of Physics, NASc KR 265a, Chui prospect, Bishkek, 720071, Kyrgyz Republic*
[4] *Atomic Energy Agency, Tajik Academy of Sciences, Dushanbe, Tajikistan*
[5] *Institute of Nuclear Physics, AS RU*

Abstract: The Navruz Project engages scientists from nuclear physics research institutes and water science institutions in the Central Asia Republics of Kazakhstan, Kyrgyzstan, Tajikistan, and Uzbekistan, and Sandia National Laboratories. The project uses standardized methods to monitor basic water quality parameters, radionuclides, and metals in the Syr Darya and Amu Darya rivers. Phase I of the project was initiated in 2000 with 15 sampling points in each of the four countries with sample analysis performed for over 100 parameters. Phase II of the project began in 2003 and expanded sampling to include at least 30 points in each country in an effort to characterize "hot spots" and to identify sources. Phase III of the project began in 2006 and will integrate decision support modeling with the existing monitoring. Overall, the project addresses four main goals: to create collaboration among Central Asian scientists and countries; to help increase capabilities in Central Asian nations for sustainable water resources management; to provide a scientific basis for supporting nuclear transparency and non-proliferation in the region; and to help reduce the threat of conflict in Central Asia over water resources. Contamination of these rivers is a result of growing population, urbanization, and agricultural activities, as well as radioactive contamination from a legacy of uranium mining and related activities of the former Soviet Union. The project focuses on waterborne radionuclides and metals because of the importance of these contaminants to public health and political stability in Central Asia.

Keywords: sustainable water resources management, radionuclides and metals, Syr Darya and Amu Darya River

1. Introduction

Political, cultural, and economic stability in Central Asia requires a reliable and sustainable supply of high-quality fresh water for agriculture and direct human consumption. Fresh water in Central Asia is provided on the surface by an international, transboundary river system, and so the most effective management of the resource will result from a collaborative, international, transboundary approach. Freshwater resource management requires long-term, basin-wide data sets shared among all transboundary partners and integration with decision support systems that allow access by a wide range of users, including scientists, policy makers, and the public.

Central Asia's history of extensive uranium mining and fabrication highlights its high potential for radionuclide and metals pollution in regional river systems. Both radionuclides and metals pose a serious long-term public health risk in the region. Data collected as part of the first two phases of the Navruz Project (2000–2006) show significant radioactive contamination levels at localized points in the region, due primarily to the Soviet-era legacy of uranium mining and waste processing. These contaminants represent a significant threat to public health and regional security, since natural events (such as heavy rainfall and flooding) or terrorist activities could result in the accidental or intentional movement of radioactive materials into public water supply systems. Interestingly, results from across the basin do not indicate widespread, serious contamination problems as many researchers expected.

The monitoring and data sharing system developed in the Navruz Project provides both baseline data and the international scientific infrastructure for tracking and addressing waterborne radionuclide contamination, however it is propagated. Preliminary results from the Navruz Project demonstrate the need for expanded joint scientific studies to more completely understand fate and transport of all kinds of contaminants.

The data from the project, along with numerous documents describing the project and some publications that have come from the project, are available to all at the following website: https://waterportal.sandia.gov-/centasia.

Founding partner institutions in the Navruz Project include the Institute of Nuclear Physics in Almaty, Kazakhstan; the Institute of Physics in Bishkek, Kyrgyzstan; the Atomic Energy Agency of the Academy of Sciences in Dushanbe, Tajikistan; the Institute of Nuclear Physics in Ulugbek, Uzbekistan; the Center for Non-proliferation Studies of the Monterey Institute of International Studies; and Sandia National Laboratories Cooperative Monitoring Center. Current efforts are aimed at engaging other institutions throughout Central Asia [1–3].

2. Methods

One of the great obstacles in transboundary river studies and transboundary river management has been the different technical methods for sampling and data collection used by different riparian nations and institutions, often rendering the resulting data poorly comparable across transboundary basins. Planning for the Navruz Project began in 1999 and was quickly followed in 2000 with a meeting of project partners in Uzbekistan to standardize methods for field data collection, sampling, sample handling, sample analysis, and data sharing. Those efforts culminated in a collaboratively developed, transboundary sampling and analysis plan [4], which is available at the website shown above.

Sampling and data collection began in 2000. Three sampling events occurred as part of Navruz 1, in the fall, spring and fall of 2000–2001. The monitoring occurred at 15 sampling sites in each country (for a total of 60 across the basin) on the main stems and the major tributaries of the Amu Darya and Syr Darya rivers. A map showing the approximate locations of the original 15 sampling locations is shown in Fig. 1. Actual sampling locations are shown in the Appendix, and are recorded in Passell et al. (2002) [5–6]. Basic water quality parameters (Table I) were collected using identical water quality data collection instruments manufactured by Hydrolab, Inc. Radionuclide and metals data (Tables II and III) were generated using a variety of radioanalytical methods, quality assurance and control approaches, and other collaborative efforts described elsewhere [1, 4]. Analytical instruments included alpha and beta counters, gamma spectrometers, activation analysis (AA), and atom emission spectrometry with inductive-coupled plasma (AES-ICP). Sampling media included water, water dissolve, bottom sediments, aquatic vegetation, and nearby soils. Data were collected on more than 100 parameters over those five environmental media.

Figure 1. Map of Central Asia. Orange dots represent approximate locations of original sampling locations for the Navruz Project

TABLE I. Basic water quality parameters

Time	Discharge (m³/s)	Water temp. (°C)	Dissolved oxygen (% saturation)	Dissolved oxygen (mg/L)	Specific conductivity (mS/cm)	Salinity (g/L)	TDS (g/L)	Depth (m)	pH	Redox potential (mV)

TABLE II. Radionuclide parameters

Alpha activity		Beta activity							
Be-7	Na-22	Na-24	K-40	Cr-51	Mn-52	Mn-54	Co-56	Co-57	Ni-57
Co-58	Fe-59	Co-60	Zn-65	Sr-85	Y-88	Zr-95	Mo-99	Ru-103	Ru-106
Ag-108	Ag-110	Sn-113	Cd-115	Sb-122	Sb-124	Sb-125	I-131	Ba-133	Cs-134
Cs-137	Ce-139	Ce-141	Ce-144	Nd-147	Eu-152	Gd-153	Eu-154	Eu-155	Ta-182
Ta-183	Ir-192	Tl-201	Hg-203	Tl-207	Tl-208	Pb-210	Pb-211	Bi-212	Pb-212
Bi-214	Pb-214	Rn-219	Ra-223	Ra-224	Ra-226	Th-227	Ac-228	Ra-228	Th-228
Th-229	Pa-231	Th-231	Th-232	Pa-233	U-235	Np-237	U-238	Pu-239	Am-241
Cm-243									

Alpha and beta activity reported in Becquerel/kilogram (Bq/kg)
Water (dissolved) and water (suspended) reported in Becquerel/liter (Bq/L)
Sediments, vegetation, and soils reported in Bq/kg.

TABLE III. Metals parameters

Al	Ag	As	B	Ba	Be	Ca	Cd	Co	Cr	Cu	Fe	Hg
K	Mg	Mn	Mo	Na	Ni	Pb	Se	Si	Tl	V	Zn	

Water (dissolved) and water (suspended) reported in micrograms/liter (µg/L). Sediments, vegetation, and soils reported in µg/g.

One of the goals of the Navruz Project has been to develop a long-term data set for rivers of the Aral Sea Basin. In Navruz 2, seasonal data and sample collection and analysis continued at the original 15 locations in each country and were expanded to include a second 15 locations in each country. The second 15 locations were chosen by each partner institution to help characterize "hot spots", or areas of high radioactivities, and to help identify the sources of those hot spots.

In Navruz 3, data and sample collection and analysis continued at the original 15 locations as a way of continuing the development of the long-term data set. Sampling and data collection also continued at the second 15 locations, although some of those locations changed as understanding of the hot spots and sources evolved.

Navruz 3 also initiated transboundary collaboration among the partners on system dynamics decision support modeling in the Aral Sea Basin. This modeling approach, tested in various settings in the U.S. and Asia [4, 7], employs system dynamics models with an interactive interface. The interactive capability – meaning that users can change variables, run the model, and view output in real time – makes this approach applicable to resource management, public education, policy making, and legislation, by both scientists and non-scientists. Models are built and run on laptops, variables can be set and reset by users, and output can be generated in seconds to minutes, allowing real time assessment of different management strategies and scenarios. Modeling efforts as part of Navruz 3 are just beginning as this writing takes place.

A weakness in the existing Navruz Project has been the absence of consistent river discharge data. This absence has been due to the physical difficulties and expenses associated with collecting these data. Gidromet agencies in each of the Central Asian countries once collected regular discharge data and still have much of the capability to do so, but since the collapse of the Soviet Union the capabilities and activities of these agencies have decreased. Navruz 3 will also broaden the current Central Asia collaboration to include Gidromet agencies in each of the countries.

3. Results

Data collected as part of the Navruz Project are available to partner scientists and to the public at this website: https://waterportal.sandia.gov/centasia.

Several publications have described the results of sample and data analysis from Navruz 1 [1–6, 8–9]. Detailed graphs, tables, and other analyses of early data can be found in those publications. Results from work under Navruz 2 are described in subsequent chapters in this volume.

In general, early results showed elevated levels of some naturally occurring and anthropogenic radionuclides at some locations throughout the Aral Sea basin. However, the widespread occurrence of high concentrations of radionuclide contamination throughout the basin has not appeared as expected by many workers. These high concentrations were expected because of the Soviet-era legacy of mining and fabrication of radionuclides and nuclear weapons testing in the former Soviet republics of Central Asia. Numerous metals concentrations across the basin did show elevated levels, likely as a result of extensive historic and current mining operations in the basin.

The largest concentration of natural radionuclides in bottom sediments occurred in the eastern part of the Chardarya reservoir of eastern Kazakhstan. Increased concentrations of all natural radionuclides and uranium also occur near the city of Shieli, Kazakhstan.

In Uzbekistan, the highest concentrations of natural radionuclides in soil and bottom sediments were found in samples from the Akhangaran River near the cities of Yangiabad, Angren, and Tuyabuguz. The Akhangaran River is a tributary of the Syr Darya in the Ferghana Valley, not far from Tashkent. In these samples, the concentration of the radionuclides of the ^{238}U and ^{232}Th families frequently exceeded 100 Bq/kg. Alpha and beta measurements from vegetation also showed significantly increased activity in those locations.

In Tajikistan, the highest concentration of natural radionuclides in the ^{238}U and ^{232}Th families is found on the Syr Darya near Khudzhand, also in the Ferghana Valley. Additionally, the highest contamination by ^{137}Cs in the region was found in the rivers that are tributaries to the Amu Darya and that are fed from the glaciers from the Pamir Mountains. The sources of this contamination require further investigation.

As an upstream country on the Syr Darya, Kyrgyzstan enjoys relatively low levels of contamination by radionuclides and metals compared with its downstream neighbors. However, considerably higher levels of radionuclide contamination were found downstream from the Mailuu Su uranium tailings, compared to levels immediately upstream. The Mailuu-Su River is a tributary to the Syr Darya near the Ferghana Valley.

The basic water quality data indicated a number of notable trends. There was a strong upward trend in the mean specific conductivity of the river water. These levels increased from less than 500 μS/cm in the upper reaches to over 1,500 μS/cm in the lower reaches. The highest specific conductivity occurred at Chinaz in Uzbekistan. The pH content of the river was between 7.5 and 8.3 throughout the length of the river. There were notable increases in pH at the Toktogul Reservoir and at Chyily in Kazakhstan.

Several interesting trends for metals emerged from the data. Results reported indicate all sampling seasons unless specifically noted. Results show high concentrations of selenium at TJ-13, UZ 06, KZ 08, and KZ 12 see Appendix for actual sampling code locations.) These concentrations exceeded 0.05 μg/L, the United States Environmental Protection Agency (EPA) drinking water safety standard.

The copper content of the river increased steadily from KG 03 through KZ 15 KZ 08, 09 and 12 had copper levels in excess of 1,300 ug/L, the maximum permissible level according to the EPA. KZ 8 and 12 also had high concentrations of chromium and arsenic.

Chromium data were inconsistent. Chromium content average over all sampling events decreased along the length of the river in the water dissolved medium. Chromium concentrations measured during spring sampling events alone increased from upstream to downstream in the water dissolved medium. Chromium content increased from the 1–2 μg/L in upstream reaches to roughly 4–6 μg/L in downstream reaches. There was also a slight increase in the chromium content in the soils and bottom sediments data from upstream to downstream.

The site most consistently contaminated with heavy metals along the river basins occurred downstream of Kyzyl-Orda in Kazakhstan. The site had high levels of arsenic, chromium, copper, lead and mercury. The site showed a distinct increase across most of the metals from the sampling location located upstream of Kyzyl-Orda.

4. Conclusions

The Navruz Project is a collaborative, international, transboundary data and sample collection, data sharing, and data analysis project engaging institutes of nuclear physics and other institutions in Kazakhstan, Kyrgyzstan, Tajikistan and Uzbekistan, in the Syr Darya and Amu Darya basins of Central Asia. Overall findings show some high concentrations of some radionuclides in sites across the basin, but generally lower concentrations than were expected based upon the nuclear legacy of the region. Some metals concentrations increase from upstream to downstream, as expected as a result of mining past and current mining operations in the basin. Salinity increases from upstream to downstream, as expected.

The project represents a strong example of international, transboundary cooperation on water resources, complete with standardized approaches, mechanisms for data sharing, and collaborative transboundary data analysis. This project can be a model for transboundary water resource collaboration in other regions of the world.

5. Acknowledgements

We gratefully acknowledge the support and assistance from Sandia National Laboratories, the U.S. Department of Energy National Nuclear Security Agency's Office of Nonproliferation Policy, the International Science and Technology Center (ISTC), and the Science and Technology Center of the Ukraine (STCU), the home institutions and governments of all the Navruz Project partners, and the NATO Science for Peace Subprogramme. Sandia National Laboratories is a multiprogram laboratory operated by Sandia Corporation, a Lockheed-Martin Company, for the United States Department of Energy under contract DE-AC04-94AL85000.

References

1. Barber, D.S., Yuldashev, B.S., Kadyrzhanov, K.K., Eleukenov, D., Ben Ouagrahm, S., Solodukhin, V.P., Salikhbaev, U.S., Kist, A.A., Vasiliev, I.A., Djuraev, A.A., Betsill, J.D., Passell, H.D., Tolongutov, B.M., Pozniak, V.L., Radyuk, R.I., Alekhina, V.M., Kazachevskiy, I.V., Knyazev, B.B., Lukashenko, S.N., Khazekber, S., Zhuk, L.I., Djuraev, A.N., Vodovina, E.D. and Mamatibraimov, S. (2003). Radio-ecological situation in river basins of the Syrdarya and Amudarya of Central Asia according to the results of the project "Navruz". In: N. Birsen and K.K. Kadyrzhanov (eds.), *Environmental Protection Against Radioactive Pollution*, Kluwer, The Netherlands, pp. 39–51.
2. Barber, D.S., Betsill, J.D., Mohagheghi, A.H., Passell, H.D., Yuldashev, B., Salikhbaev, U., Djuraev, A., Vasiliev, I. and Solodukhin, V. (2005). The Navruz experiment: Cooperative monitoring for radionuclides and metals in Central Asian Transboundary Rivers. *Journal of Radioanalytical and Nuclear Chemistry* **263**(1): 213–218.
3. Kadyrhzanov, K.K., Barber, D.S., Solodukhin, V.P., Poznyak, V.LO., Kazachevskiy, I.V., KInyazev, B.B., Lukashenko, S.N., Khazhekber, S., Betsill, J.D. and Passell, H.D. (2005). Radionuclide Contamination in the Syrdarya river basin of Kazakhstan, Results of the Navruz Project. *Journal of Radioanalytical and Nuclear Chemistry* **263**(1): 197–205.
4. Passell, H., Tidwell, V. and Webb, E. (2002). Cooperative modeling: An approach for community-based water resource management. *Southwest Hydrology* **1**(4): 26.
5. Passell, H., Barber, D., Betsill, D., Littlefield, A., Mohagheghi, A., Shanks, S., Lojek, C., Yuldashev, B., Salikhbaev, U., Radyuk, R., Djuraev, A., Vasiliev, I., Tolongutov, B., Alekhina, V., Solodukhin, V. and Pozniak, V. (2002). The Navruz Project: Transboundary monitoring for radionuclides and metals in Central Asian rivers; Sampling and analysis plan and operations manual. SAND Report 2002-0484. Sandia National Laboratories, Albuquerque, NM, 87185.
6. Passell, H., Barber, D., Betsill, D., Littlefield, A., Matthews, R., Mohagheghi, A., Shanks, S., Yuldashev, B., Salikhbaev, U., Radyuk, R., Djuraev, A., Vasiliev, I., Tolongutov, B., Alekhina, V., Solodukhin, V. and Pozniak, V. (2003). The Navruz Project: Transboundary monitoring for radionuclides and metals in Central Asian rivers; Data report. SAND Report 2003-1149. Sandia National Laboratories, Albuquerque, NM, 87185.
7. Tidwell, V.C., Passell, H.D., Conrad, S.H. and Thomas, R.P. (2004). System dynamics modeling for community-based water planning: Application to the Middle Rio Grande. *Aquatic Sciences* **66**: 357–372.
8. Vasiliev, I.A., Barber, D.S., Alekhina, V.M., Mamatibraimov, S., Barber, D., Betsill, D. and Passell, H. (2005). Uranium levels in the Naryn and Mailuu-Suu rivers of Kyrgyz Republic. *Journal of Radioanalytical and Nuclear Chemistry* **263**(1): 207–212.
9. Yuldashev, B.S., Salikhbaev, U.S., Kist, A.A., Radyuk, R.I., Barber, D.S., Passell, H.D., Betsill, J.D., Matthews, R., Vodovina, E.D., Zhuk, L.I., Solodukhin, V.P., Pozniak, V.L., Vasiliev, I.A., Alekhina, V.M. and Djuraev, A.A. (2005). Radioecological monitoring of transboundary rivers of the Central Asia Region. *Journal of Radioanalytical and Nuclear Chemistry* **263**(1): 219–228.

6. Appendix: Sampling codes and actual locations

6.1. KYRGYZSTAN

KG-01, Kichi-Naryn River before the confluence with the Chong-Naryn River
KG-02, Chong-Naryn River before the confluence with the Kichi-Naryn River
KG-03, Naryn River before the confluence of the Kichi-Naryn
KG-04, At-Bashy River before its confluence into the Naryn River
KG-05, Naryn River after the confluence of the At-Bashy tributary
KG-06, Chychkan River before the confluence into the Toktogul water pool
KG-07, Naryn River before its confluence in the Toktogul reservoir (hydrological post Uch-Terek)
KG-08, Toktogul reservoir
KG-09, Naryn River after the Toktogul reservoir (region of Kara-Kul town)
KG-10, Naryn River, southeast part of the town of Tashkumyr
KG-11, Mailuu-Su River, on the bridge (boundary with Uzbekistan)
KG-12, Mailuu-Su River at the departure from Mailuu-Su town
KG-13, Mailuu-Su River near the transformer factory
KG-14, Mailuu-Su River, right tributary
KG-15, Mailuu-Su River 200 m from the tributary

6.2. KAZAKHSTAN

KZ-01, Chardarya reservoir, southeastern coast between the Keles and Kurukkeles Rivers
KZ-02, Chardarya reservoir, northeastern coast near Chardarya
KZ-03, Keles River, Saryagash town (upstream)
KZ-04, Keles River, Saryagash town, (downstream)
KZ-05, Badam River, Chymkent (upstream)
KZ-06, Arys River before the confluence with the Badam River, Chymkent (downstream), near Obruchevka
KZ-07, Syrdarya, Chernak village (below Turkestan town)
KZ-08, Syrdarya, Chyily (upstream), Tomlnaryk village
KZ-09, Syrdarya, Chyily
KZ-10, Syrdarya, Chyily (downstream), Zhulek village
KZ-11, Syrdarya, Kyzyl-Orda (upstream), Belkul village
KZ-12, Syrdarya, Kyzyl-Orda (downstream), Abaj village
KZ-13, Syrdarya, Korkyt village, below Zhusa town
KZ-14, Syrdarya, Bajkonur town, below Torwtam village, near Bay-Kozha
KZ-15, Syrdarya, Kazalinsk

6.3. TAJIKISTAN

TJ-01, Varzob River, 18 km above Dushanbe city
TJ-02, Varzob River, 9 km below Dushanbe city
TJ-03, Kafirinigan River, 1 km above the confluence with the Varzob River
TJ-04, Kafirnigan River, 3 km below its confluence with the Elok River
TJ-05, Kafirnigan River, at the Shaartuz railway bridge
TJ-06, Elok River, 1 km above its flow into the Kafirnigan River
TJ-07, Vakhsh River, at the Dzhilikul bridge
TJ-08, Vakhsh River, "Chiluchor chashma," the spring
TJ-09, Vakhsh River, 1 km below Norak City
TJ-10, Yekhsu River, at hydrological post "Vose", the settlement Vose
TJ-11, Kyzylsu River, 5 km from the settlement Vose, before its confluence with the Yekhsu River
TJ-12, Kyzylsu River, Gulistan Village
TJ-13, Syrdarya, 60 km above the Kayrakkum reservoir (unfinished frontier bridge), settlement Bulok
TJ-14, Syrdarya, the bridge on the entrance of Khudzhand city
TJ-15, Isfara River, the settlement Yangiobod between Rabot city and Nefteobod city

6.4. UZBEKISTAN

UZ-01, Amudarya, Kyzyldzhar village, Karakalpakstan, 1 km above the terminating range of the Amudarya River (the nearest town is Kungrad)
UZ-02, Amudarya, Kipchak town, Karakalpakstan, 0.5 km above the town
UZ-03, Amudarya, Tuyamuyun site, 8 km below the dam (Khorezm region)
UZ-04, Karadarya, Namangan region, 20 km southwest from Namangan, at Kol' village
UZ-05, Syrdarya, Bekabad, Tashkent region, 0.9 km below the dump of drainage waters of "Vodokanal" enterprise
UZ-06, Syrdarya, Chinaz town, Tashkent region, 3.5 km SSW from Chinaz
UZ-07, Ankangaran River, Yangiabad town, Tashkent region, 5.5 km below Dukant village
UZ-08, Akhangaran River, Angren town, Tashkent region, 5.5 km below the Angren dam
UZ-09, Akhangaran River, Tuyabuguz, Tashkent region, Soldatskoe village, 0.5 km above the outfall of the Akhangaran River
UZ-10, Chirchik River, Gazalkent town, Tashkent region, 3.5 km below the town
UZ-11, Chirchik River, Kibraj village, Tashkent region, 3 km below the UZKTZhM enterprise sewage effluent
UZ-12, Chirchik River, Tashkent City, Tashkent region, 3 km below the sewage effluent from the Segeli KSM plant
UZ-13, Zaravshan River, Ravatkhodzha, Samarkand region, 3.7 km below the outfall of the Taligulyan dump
UZ-14, Zaravshan River, Kattakurgan, Samarkand region, 0.8 km below the outfall of the Chegonak collector
UZ-15, Zaravshan River, Navoi City, Navoi region, 0.8 km below the sewage effluent from "NavoiAzot" enterprise

CHAPTER 19

SEVERAL APPROACHES TO THE SOLUTION OF WATER CONTAMINATION PROBLEMS IN TRANSBOUNDARY RIVERS CROSSING THE TERRITORY OF ARMENIA

G.M. ALEKSANYAN[1], A.N. VALYAEV[2], K.I. PYUSKYULYAN[3]
[1]*Yerevan State University, Department of Geology,*
1 Alex Manoogian Str., 0025, Yerevan, Armenia
[2]*Nuclear Safety Institute of the Russian Academy of Sciences*
(IBRAE RAS),
52 b. Tulskaya str., 115191, Moscow, Russia
[3]*Armenian Nuclear Power Plant, Medzamor, Republic of Armenia*

Abstract: Transboundary water resources are of significant importance for all the EECCA countries. The break-up of the Soviet Union and other States has created new borders and new transboundary waters. The region has now about 150 major transboundary rivers that form or cross borders between two or more countries, some 25 major transboundary and international lakes and some 100 transboundary aquifers. Many catchments drain into closed seas or into landlocked lakes and pollution transported by rivers to seas and lakes has a major influence on these ecosystems. Many industrial enterprises are situated in the basins of the rivers, and there are major concerns related to the contamination of the fresh water ecosystems. The present paper describes the project "Monitoring of the probable sources and mechanisms of pollution with radionuclides, toxic and chemically dangerous elements of the rivers Kura and Araks with their inflows on the territory of Armenia" which is an important component of the International Program "Joint international studies of the Caspian river basins' pollution on the territories of Russia, Kazakhstan, Georgia, Azerbaijan and Armenia for the transboundary control".

Keywords: water quality, radionuclides, metals, transboundary rivers, Armenia

1. Introduction

The UN Economic Commission for Europe (UNECE, January 15–16, Geneva, 2004) held a Regional Implementation Forum on Sustainable Development. As a result a document was published entitled "Water and sanitation in the UNECE region (including the countries of Eastern Europe, South Caucasus and Central Asia (EECCA countries)):

achievements in regulatory aspects, institutional arrangements and monitoring since Rio, trends and challenges", where the following was summarized:

- Water shortages, unsafe water and inadequate sanitation are reported on every continent, including Europe. Worldwide, some 2.4 billion people lack access to basic sanitation and 1.2 billion, or one in five, lack safe drinking water.
- A number of European countries abstract at least as much surface water as they generate. Twenty European countries depend on water coming from neighboring countries: more than 10% of their waters are formed abroad. For five countries, 75% of their resources are formed in upstream countries. Reasonable and equitable use of transboundary waters (and watersharing between different sectors in the national economy and between countries) is therefore a particular challenge.
- Owing to the essential transboundary nature of waters in the UNECE region, hydroeconomic activity, water management and sanitation-related issues could become a source of disputes between countries. This requires new policies that are all-embracing and environmentally sound, and that involve the public at large.

Transboundary water resources are of significant importance for all the EECCA countries as a whole. The break-up of the Soviet Union and other States has created new borders and new transboundary waters. The region now has about 150 major transboundary rivers that form or cross borders between two or more countries, some 25 major transboundary and international lakes and some 100 transboundary aquifers. Many catchments drain into closed seas or into landlocked lakes and pollution transported by rivers to seas and lakes has a major influence on these ecosystems.

In many EECCA countries and in South-East Europe, hydrometric and water-quality monitoring networks are deficient. The former Soviet Union had a relatively

Figure 1. The Caspian region

well-established water-monitoring network, but after its break-up and the independence of the EECCA countries, this infrastructure started to degrade with the deterioration of the States' economies.

Another problem is the "data-rich but information-poor" syndrome. Many countries with planned economies used to collect and store huge amounts of water-monitoring data, but did not translate the data into useful, policy-relevant information. There is still a general lack of environmental monitoring and comparable data and information on water quality and quantity in many EECCA countries. National surface-water monitorring systems are not coherent, as neither the data reporting systems nor the methodologies are harmonized.

A joint proposal entitled "Radiation study and monitoring of the Caspian Sea rivers" was elaborated and presented to ISTC (International Science and Technology Centre) competition for funding support by five countries of the Caspian basin (Russia, Georgia, Armenia, Azerbaijan and Kazakhstan). One objective of the project was to create a well-grounded system and a scheme of radiation and hydrochemical monitoring and also database on pollution of the river basins of the Caspian region (Fig. 1). The Armenian research team will implement the ISTC project A-1311 entitled "Study of the level of pollution with radionuclides, toxic and chemical hazardous elements and creation of monitoring system of Kura and Araks basins in Armenia" (Fig. 2).

2. The Project Objective

The objective of this Project is the study and the characterization of pollution with radioactive nuclides, toxic and chemical hazardous elements (Table I), the creation of well-grounded system and scheme of radiation and hydrochemical monitoring of the basins of Caspian Sea rivers (Kura and Araks) on the territory of the Republic of Armenia. In the course of the Project implementation, verification of the available experimental data and conducting of additional investigations within reference test areas will be performed in order to ensure compatibility and agreement of the data, improve the methodology of data collection and improve the quality of primary information used when developing the routines for primary information collection and processing and also for developing the radionuclide migration models.

The Kura and Araks rivers are the life supporting arteries for the large region of the Southern Caucasus. The basins of these rivers include the following states: Georgia, Armenia, Azerbaijan, Turkey and Iran. There is serious concern about the epidemiological situation in the region, particularly in the midstream and lower stream of the rivers. One of the main reasons for the high sickness rate of the population is attributed to significant pollution of the rivers and the tensed environmental and radioecological conditions in their basins. Such a situation appeared due to the intensive (diverse and uncontrolled) industrial and economical activity, and also in virtue of the geological and geochemical features of the region (Fig. 3).

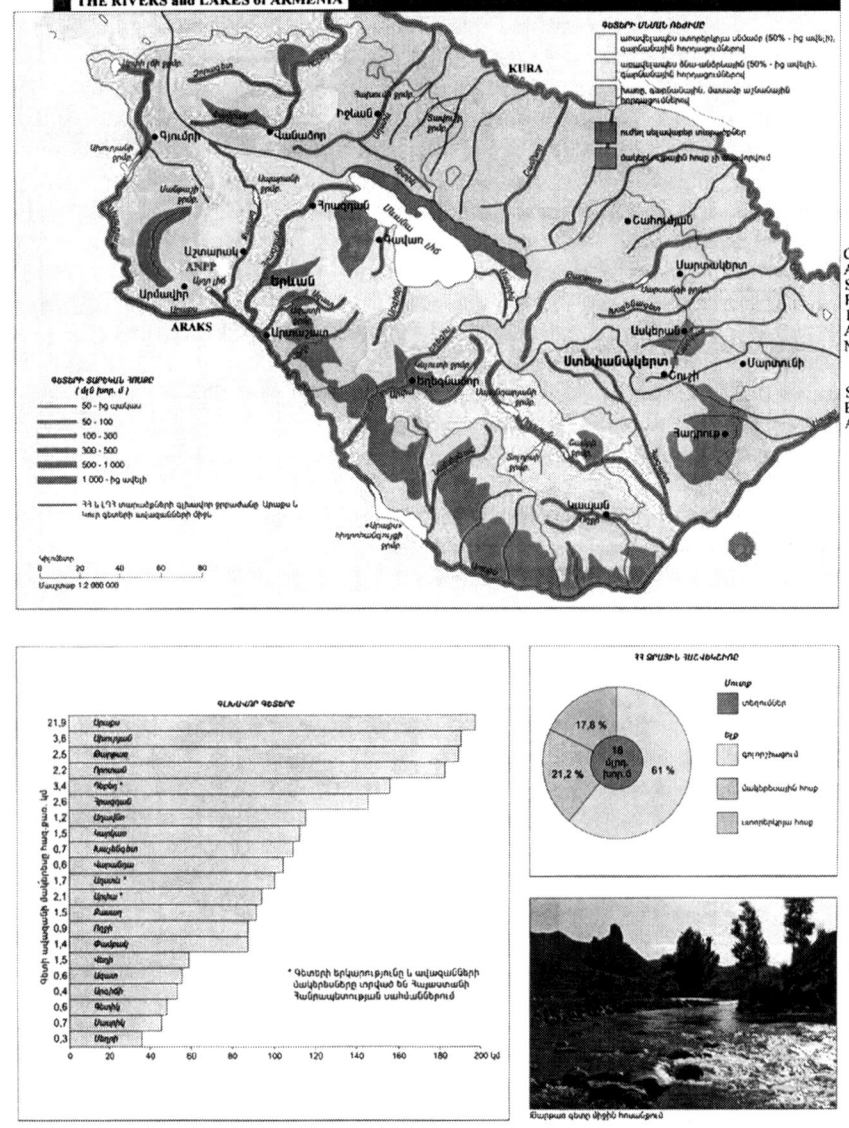

Figure 2. Kura and Araks basins in Armenia

For the Araks and Kura river basins in Armenia a high pollution rate with natural radionuclides (series of ^{238}U, ^{235}U, ^{232}Th) is typical. Many compartments in the environment, besides the radionuclides, are characterized by the large concentration of the heavy metals and toxic elements (Cr, Pb, Zn, Cu, Mo, As, Se, Te, Fe, V, Co,

Mn, Ni, Cd, Sb, Hg, Th, U). A region, adjoining to the Zangezur, Bargusht, Urc, Shirak, Pambak, Bazum and Sevan mountain ridges and to their spurs, is especially distinguished by this feature.

The hydrochemical analysis of the water of the rivers and the reservoirs polluted with the sewage on the territory of Armenia, includes the following parameters of

TABLE I. The list of indices liable to investigation while the assessment of open reservoir water quality in Republic of Armenia (Araks river, its inflows and the inflows of Kura river)

The group of indices under determination	Names of indices
1	2
1) Generalized indices:	Hydrogen index (pH) Surfactant species (anionic), oxidability The level of pollution with natural radioactive nuclides (series ^{238}U, ^{235}U, ^{232}Th)
2) Organic chemical agents:	• DDT – dichlorodiphenyltrichloroethane- dichlorophenyl 3 chloro-ethane, its metabolites • DDE – dichlorodiphenyl dichloroethylene • DDD – dichlorodiphenyl dichloromethylmethane • HCCH - hexachlorocyclohexane its isomers (α, β, J) • J – isomer – Lindane (pesticide) • L and β isomers • Atrazine • Simazine • 2,4 D- phenoxyacetic acid – Phenols volatile (summarized)
3) Complex indices of toxicity:	• The sum of nitric oxid NO_2 NO_3 • The presence of heavy metals and toxic elements (Cr, Pb, Zn, Cu, Mo, As, Se, Te, Fe, V, Co, Mn, Ni, Cd, Sb, Hg, Th, U)
4) Indices of epidemiological safety Bacteriological: Virological: Parasitologic:	The total microbe number (TMN) under the temperature of $22°C$ and $37°C$ • General coliform bacteria • Thermotolerant coliform bacteria • Spores of sulfitoreduced clostridials • Glucose-positive coliform bacteria • Pseudomonas aeriginosa • Coliphages • Oocysts of cryptosporidia • Cysts of lamblia • Helminth ovums

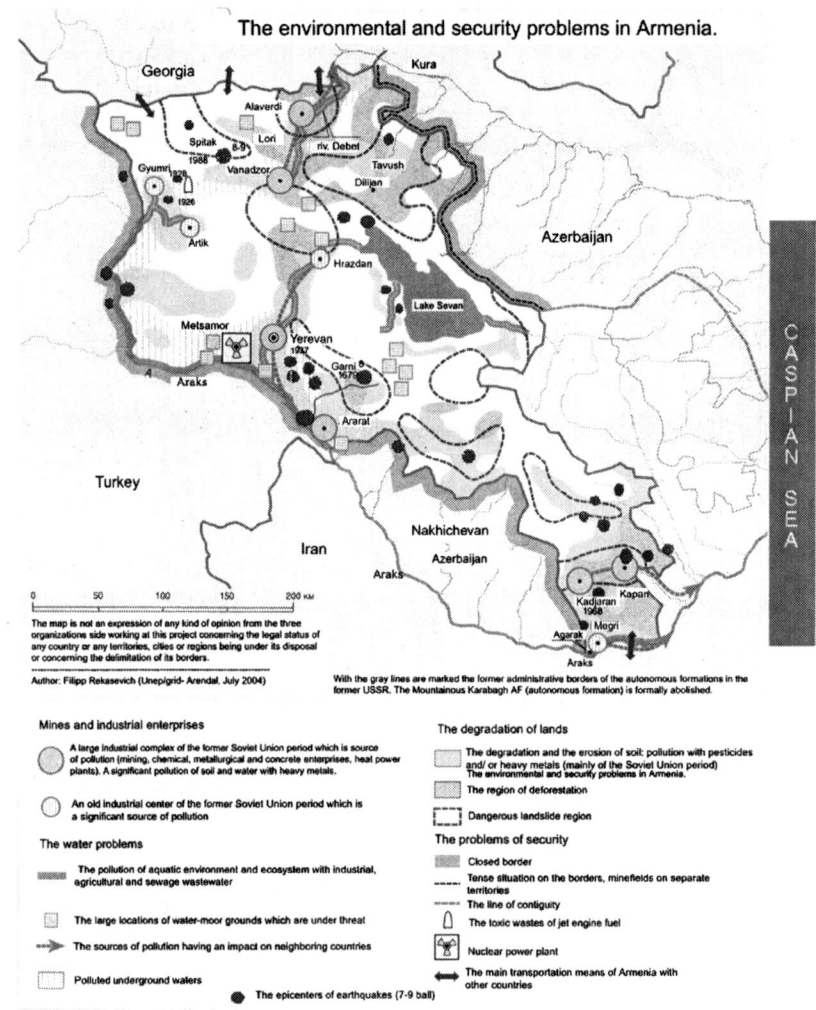

Figure 3. Geological and geochemical features of the region

pollution (in mg/l): oxidability (the dissolved oxygen O_2), magnesium (Mg^{2+}), chlorides (Cl^+), sulfates (SO_4^2), a mineralization, bichromatic oxidability, iron (Fe), ammonium nitrogen (NH^{4+}), nitrate nitrogen (NO^{3-}), the suspended substances, biological need of oxygen (BNO), zinc (Zn), copper (Cu) and other metals, mineral oil, phenols, synthetic surface-active substances (SAS), active (pH) reaction and temperature of water. It is well known that one of the pollution quantitative characteristics of water is the content of the ammoniac ions. Nitrogen containing substances (NH_3), anhydrites of the nitric NO_2 and nitrogenous N_2O_3 acids are formed in water mainly as a result of decomposition of the compounds in the sewage.

Sometimes, the ammonia present in the water can be of inorganic origin, resulted by formation after restoration of nitrates and nitrites by humic substances, hydrogen sulphide, oxide iron, etc.

Many industrial enterprises are situated in the basins of the rivers (Akhurian, Sevdjur-Kasakh, the basin of Sevan Lake (28 rivers and inflows flow into the lake) Hrazdan, Azat-Garni, Arpa, Meghriget, Vokhji and Vorotan, flowing into Araks, the long-lasting exploitation of which added to the pollution of waters of Araks river; heavy toxic metals, such as: Cr, Zn, As, Se, Cd, Sb, Hg, Th, U. Many of these varuiables, placed in "Hydro-chemical bulletins" of Hydrometeoservice of Armenia, were determined approximately, the obtained data have many faults and they can't serve as a basis for scientific summarizing.

In the same region Armenian NPP is situated. For the period of its exploitation (exploitation began in 1976, after Spitak earthquake in 1989 the station was conserved and then it was set on again in 1995) it accumulated large amounts of highly active, medium-active and low-active wastes. During the exploitation period, the ANPP has accumulated and kept in corresponding storages the following quantity of SRW (solid radioactive wastes): low-active – 5,275.5 m^3, middle-active – 931.2 m^3, highly active – 29.87 m^3. Water disposals from the Armenian NPP at last enter the river Araks. Besides waste-waters, the Armenian NPP provides via emissions to the atmosphere a certain amount of radionuclides ("Report about the radiation atmosphere in sanitary-defense zone and zone of observation of Armenian NPP Metzamor, 1978–2002").

It is also worthy to mention that just next to Armenian NPP is situated the object CJS "Sterilization of radioactive wastes". This enterprise is also situated in the basin of river Sev-Jur, entering river Araks. The square of this object is situated on the territory where the ground waters pass, at 25 m depth, with a direction towards the river Sev-Jur.

Besides, fallout from the accident at Chernobyl NPP did also reach the territory of Armenia. In particular, the Cs-137 concentration in the atmosphere for the period from May–August 1986 increased with a factor of 100, and the concentration in soil increased with a factor of 2 ("Report about the radiation atmosphere in sanitary-defense zone and zone of observation of Armenian NPP Metzamor, 1986, 1987). That's why it will be of special interest to investigate further consequences of the fallout.

It is a necessity to continue the work on revealing other pollution sources, more detailed studies of the features of these sources and mechanisms of input of all kinds of the pollutants in the basic life supporting fresh-water arteries – the rivers Kura and Araks, and as a result, a decision making on prevention, or restriction of this negative process is needed.

The successful solving of the ecological problems is not possible within a separate country and should include the joint and coordinated efforts of all the countries. The present Project "Monitoring of the probable sources and mechanisms of pollution with radionuclides, toxic and chemically dangerous elements of the rivers Kura and Araks with their inflows on the territory of Armenia" is an important component of the International Program "Joint international studies of the Caspian river basins' pollution on the territories of Russia, Kazakhstan, Georgia, Azerbaijan and Armenia for the transboundary control".

3. Expected Results and Their Application

Expected results from the Project implementation will be:
- New data on the pollution with radionuclides and toxic elements of the rivers Araks and Kura basins.
- Identification of the most polluted sites and objects.
- Characterization of the most polluted sites and objects, representing potential danger of further emission of the pollutants.

The final result of the present Project will be the creation of well-grounded system and scheme for radiation and hydrochemical monitoring and also a database on the pollution of the basins of the Caspian Sea rivers (Araks and Kura) on the territory of the Republic of Armenia.

4. Competence of the Project Participants

Highly skilled specialists will participate in the project work (2 – doctors of sciences, 13 – candidates of sciences and 9 – engineers): physicists, chemists, engineers and technicians, experienced in the field of nuclear physics methods of the analysis, radiochemistry, atomic absorption analysis, mass-spectrometry, applied geochemistry, nuclear techniques, and also programmers, experts on the data processing, geoinformation technologies and on the mathematical simulations.

5. Scope of Activities

Review of the available data and potential, choice of checkpoints and preparation of concrete research programmes;
- A retrospective review and the critical analysis of all kinds of activities and the natural phenomena having effect on the ecological and radio-ecological situation.
- Selection of sites and objects for research based on the available information.
- Development of the methodical approach for the research.
- Field works (twice a year: in spring and autumn).
- Radiometric (γ) spectrometric measurements.
- Definition of the coordinates of the points for measurements and sampling.
- Regular sampling of the water.
- Sampling of the soil, bottom sediments and plants.
- Conducting measurements determining flows, and also basic qualitative and quantitative physic-chemical parameters of the water content.

Laboratory researches (including analysis of the samples, selected according to the International Project, on the territory of Armenia, Georgia and Azerbaijan – cooperation on a contract basis)

- Preparation of the samples for all types of analyses, depending on their features.
- Chemical, radiochemical, etc. types of the physical and chemical analysis of the samples of soil, bottom sediments, water and plants.
- Carrying out an alpha- beta- gamma- (α-, β-, γ-) spectrometric measurements.
- Mass-spectrometer study of the isotopes' content of uranium and others radionuclides in the samples of soil and bottom sediments.

Processing and interpretation of the results

- The overview of the available data and potential. The choice of checkpoints. Preparation of concrete research programs.
- The choice of sites and objects for the research (not less than 15 for each expedition) on the basis of the existing information.
- Characterization of the radionuclides and toxic elements localization in the most polluted sites and platforms of the objects.
- Development of recommendations on the control, and also on the restriction and prevention of all possible mechanisms of the pollutants emission from these sites and objects.
- Creation of the database on the radionuclides and toxic elements of the rivers Araks, Kura and their inflows, separately, on the most polluted sites and objects, and on the qualitative and quantitative parameters of the water.
- Presentation of the annual and final reports.
- Presentation of the results studies at the International, Republican and interdepartmental conferences, seminars, meetings, and their publication in the scientific journals.

6. Tasks

- Large-scale study of probable pollution with radionuclides and toxic elements of the basins of rivers Kura and Araks including their inflows on the territory of Armenia.
- The general characterization of the pollution of Araks and Kura river basins and creation of final radiation and hydrochemical monitoring system on its basis.
- Creation of a database on the quality and quantity of the water, and on the pollution rate of separate sites and objects in the neighborhood of the rivers Kura and Araks and their inflows on the territory of Armenia.

The project includes the following aspects:

- Scientists from the five participating countries will work together to establish standard data collection, sampling and sample analysis methods, including use of Hydrolab instruments for making field measurements of basic water quality

parameters. Standardization of methods will make data from all countries comparable across all watersheds and international borders.
- A set of permanent sampling locations will be identified in each country and then sampling will occur at those locations throughout the project, with the intention of developing a consistent, long-term database. Other locations may be added later in the project by individual countries. Sampling events will occur twice annually – once in the spring, and once in the fall.
- Scientists from the five participating countries will collect data on the same list of parameters. This list will include basic water quality parameters, metals, and radionuclides. Other parameters (organics, nutrients, petrochemicals, or others) may be added by individual countries. Sampling media will include water (dissolved), water (suspended), and bottom sediments. Other media may be added by individual countries.
- Scientists from the five participating countries will collaborate with other regional institutions (for example, Gidromets), where available, for acquiring river discharge data.
- A shared, internet-based database will be developed for storing and sharing project data among all partners and the public.
- Scientists from the five countries will collaborate as appropriate with scientists from the NATO Science for Peace project "Cooperative River Monitoring among Armenia, Azerbaijan, Georgia and the U.S." This collaboration will be designed to avoid redundancies between projects and to add strength and value to both projects.
- Scientists from the five countries will meet annually to review project results, resolve problems, and plan future efforts.
- As a result of the Project implementation, a database will be created on pollution with radioactive nuclides and toxic elements in the rivers Araks, Kura and their inflows on the territory of the Republic of Armenia. The database will also include the investigation of more polluted sites.
- The database created as a result of monitoring investigations will contribute to form GIS (geographical informational system) control of the ecological status of the Caspian Sea basin rivers on the territory of Armenia. The database and GIS control of the ecological status of the rivers in the Caspian Sea basin on the territory of Armenia can be provided on a commercial basis to interested organizations.

7. Demand for Results

The interest in the potential the project results has been expressed not only by the regional and international ecological and monitoring organizations, organization of emergency situations protection, health protection, strategic planning, transboundary control, the freshwater preservation etc., but also, first of all, by the specialized state organization as the Republic of Armenia has signed and ratified a number of international conventions and protocols (The framework convention on the climate change – the Kyoto protocol (Kyoto, 1997, New York); The convention on the assessment of the influence on the environment in the transboundary context

(Espo, 1992); The convention on the transboundary influence of the industrial accidents (Helsinki, 1992); The convention on the preservation and usage of transboundary waterways and international lakes (Helsinki, 1992); The protocol on the water and health (London, 1999); The Stockholm convention on the persistent organic pollutants (Stockholm, 2001) etc.), and also "The program of National nature-conservative measures" were elaborated by the Ministry of Nature Preservation of RA and approved by the RA Government.

8. Additional Developments

The completion of the Project and the creation of database on the results of monitorring of the radiation and chemical pollution of the Araks and Kura river basins should provide:

- The creation of computer hydrochemical map of pollution of the examined territories.
- The preparedness and presentation of the information in a more comprehensible way for the production managers, for the authority of ecological control and also for the public to assist the work in the sphere of steady development.
- The creation of the useful instruments of planning, which can help to assess visually the consequences and also the range of resources under disposal for solving the problems and overcoming the ecological crises and to minimize their consequences.

9. Conclusion

During the last 20 years, an especially difficult social-economic situation has developed in the Republic of Armenia. The devastating earthquake in 1988 and the further political occurrences led to a change from bad to worse in the economic and ecological situation. It's obvious that in parallel to the activity to improve the economics, it is also necessary to elaborate an idea and a program for the steady development of the republic, taking into account the international experience. The implementation of the Project will promote the steady development of Armenia as a part of the Southern Caucasus and countries situated in the Kura and Araks river basins, flowing into Caspian Sea, by using the results of radiation and chemical monitoring as an instrument to reduce the risks associated with different disasters and man-caused and ecological catastrophes connected with them.

CHAPTER 20

STUDY OF THE ECOLOGICAL STATE OF RIVERS KURA, ARAKS AND SAMUR ON AZERBAIJAN TERRITORY

R.F. MAMEDOV, G. MAGERRAMOV
Geology Institute of Azerbaijan National Academy of Sciences, 29a H.Javid av., AZ1143, Baku, Azerbaijan

Abstract: The most important freshwater bodies in Azerbaijan are the transboundary Kura, Araks and Samur rivers. The catchment areas of these rivers cover large territories of the South Caucasus and are important resources of freshwater. From year to year the water quantity in rivers reduces gradually and the water quality gets worse. There is a need to identify and characterize the sources that contribute to the contamination of the rivers flowing into the Caspian Sea, including the Kura, Araks and Samur rivers.

Keywords: water quality, contamination, transboundary rivers, Azerbaijan

1. Introduction

There is a range of existing and potential sources of contamination within the catchment area of the Caspian Sea, including poorly controlled industry. The situation became worse after the collapse of the USSR and the establishment of independent states. At present there is a need for an environmental monitoring programme in the region. It will provide a fulfilment of the regulations accepted in 1995 through the Global programmme of actions on marine environment protection against the pollution as a result of landbased activities. This international programme was agreed upon in order to supplement the part of the U.N.O. Convention on maritime law (1982) that deals with implementation of regulations of marine environment pollution from landbased sources. An important part of this programme will be to characterize the sources and transport mechanisms of contaminants in the transboundary Kura, Araks and Samur rivers, which are flowing through Georgia, Armenia and then Azerbaijan before discharging into the Caspian Sea. Four countries (Russia, Kazakhstan, Iran and Turkmenistan) border to the Caspian Sea.

The aim of the present project is to characterize the contamination, including radioactivity, in the rivers of the catchment area of the Caspian Sea including Kura, Araks, and Samur rivers running within the territory of Azerbaijan. Furthermore, a monitoring programme is to be designed and implemented. The catchment areas of the rivers in question cover a large area of the South Caucasus and represent the major freshwater resources. Incoming of organic pollutants and especially oil products and toxic substances leads to reduction of viability of the biological community.

It is necessary to mention that more than half of the territories of the South Caucasus states have country inclination towards the Caspian Sea. Thus, all the negative anthropogenic influences upon the river basin in neighbour states will be further reflected in Azerbaijan and in the Caspian Sea accordingly. Azerbaijan water resources make up about 30.9 km^3, from which about 20.6 km^3 come with transit rivers from the neighbour states [1]. So, one can say that nearly 70% of the water and, accordingly, pollutants, come to Azerbaijan from the territories of other countries, mainly from Georgia and Armenia.

According to official data for 1990 in Armenia about 2.6 million cubic metres of various wastes drain to the basin of the Araks River, and 4.6 million cubic metres to the basin of the Kura River in Georgia, From year to year the water quantity in the transit rivers reduces gradually and the quality gets worse. Incoming of organic pollutants and especially oil products and toxic substances brings about activation of reduction processes in the sediments. With the participation of organic substances, a change in the sulphur and the carbon cycle occurs resulting in oxygen deficit water and eutrophication, i.e. the basin is overgrown by water-plants, and the viability of the biological community is reduced. So, there is a necessity to carry out the work to identify the sources of pollutants and study in detail the peculiarities of the sources as well as mechanisms of the transport of all kinds of pollutants into the basic rivers flowing into the Caspian Sea, including the Kura, Araks, Samur rivers [2–7].

2. Outline of the Project

For Azerbaijan this problem is of special significance. These rivers are the main sources of potable water and in addition they are important in terms of irrigation, agriculture and fishing. Two water pipelines provide Baku city with water from the Kura River. The cities of Baku and Sumgayit as well as the settlements and villages of Absheron peninsula receive water from the Samur River, via the Samur-Divichi canal and the Jeyranbatan reservoir through two pipelines (Fig. 1).

The purification plants on the branches of the pipelines operate close to their load limits. Flooding occurs regularly along the Kura River inundating great territories, villages, settlements and even the city of Salyan (Fig. 2). The present project also includes an investigation of this territory. A database with data obtained from the present project (including the monitoring) will be provided, which should facilitate the management of these water resources.

The project include the following tasks:

- A wide-range, large-scale study determining the levels of contaminants including radionuclides in water and bottom sediments of the Kura, Araks and Samur Rivers.
- To provide information on the location and situation of the most polluted areas based on obtained data.
- To study the transport mechanism of the contaminants in question.

- To develop and implement a monitoring programme with respect to chemical contaminants including radionuclides in the Kura, Araks, Samur Rivers on Azerbaijan territory.
- Develop a database with information on the quality and quantity of river catchment water as well as characteristics of identified sources of contamination.

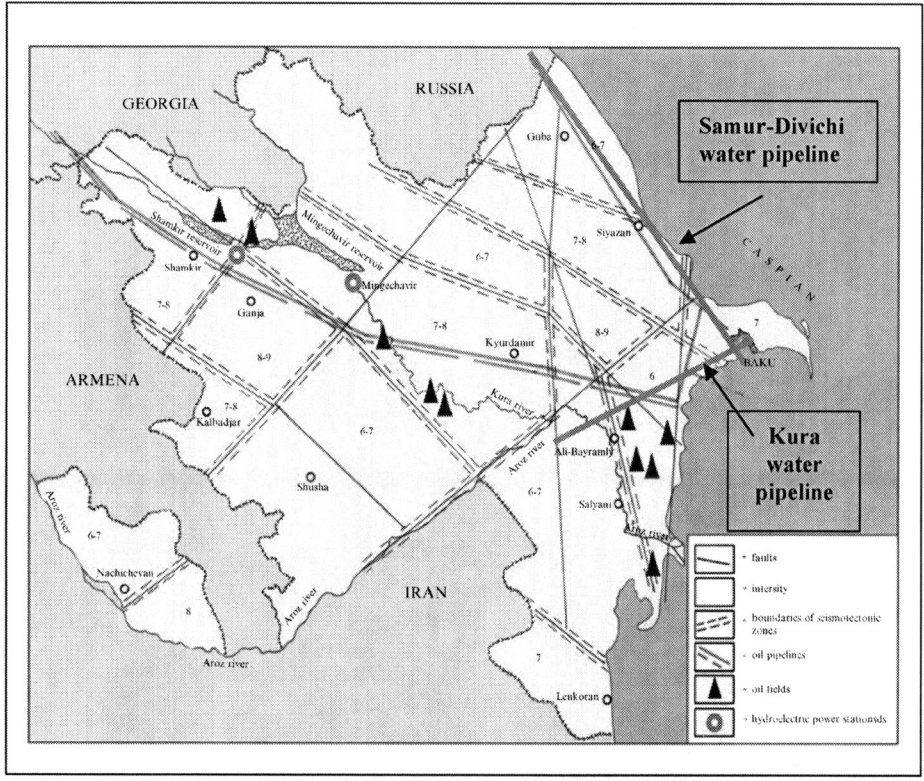

Figure 1. Map of Azerbaijan, indicating the water pipelines to Baku city

To implement the planned investigations some representative areas (stations) for study will be selected based on retrospective review and critical analysis of all activities and natural phenomena affecting the ecological situation in the region. Sampling of river water and bottom sediments will be done twice a year (spring and autumn), covering both high water flow events caused by snow melting and normal level situations mainly due to ground water inflow. Radiometric measurements and measurements determining basic qualitative and quantitative water parameters will be carried out at well defined sampling locations. The methods of chemical, hydrological, spectral, radiometric analysis as well as method of gamma-spectrometry will be applied. The obtained results should facilitate the transboundary management of the water resources within the Caspian Sea Basin.

Figure 2. Flooded areas in Salyan city

3. Conclusions

At present there are some data available on pollution of the aquatic environment of the Kura, Araks and Samur rivers. At the same time there is an absolutely obvious necessity to perform further comprehensive investigations to identify and characterise the key sources of contamination to the Caspian Sea basin. For this purpose it is necessary to:

- Investigate on a large scale the chemical, elemental and radionuclide composition of water and bottom sediments of the Kura, Araks and Samur rivers.
- Identify the most polluted areas with detailed study of the situation based on the obtained data.
- Study the contaminant transport processes.
- Design a monitoring programme for the Kura, Araks, Samur rivers on the territory of Azerbaijan.

Produce a database with information on water quality and quantity as well as on the degree and character of pollution at contaminated sites located within the river catchment areas.

References

1. Mammadov, V.A. (2003) – Water reservoirs on Kura River. Baku, R.N. Novruz – 94, 65 p.
2. Rustamov, S.G., Kashkai, R.M. – Water resources of Azerbaijan SSR. Baku, 1989, p. 181 (published in Russian).
3. State of the environment of Azerbaijan Republic. Ministry of Ecology and Natural Resources, Baku, 1997, p. 95 (published in Russian).
4. Water resources of the Transcaucasia. Edited by G.G. Svanidze, Leningrad, Gidrometeoizdat, 1988, p. 264 (published in Russian)
5. Khalilov Sh.B. (2003) – Reservoirs of Azerbaijan and their ecological problems. Baku, p. 310.
6. Mansurov A.E., Salmanov M.A. (1996) – Environment of Kura river and its reservoirs' basins. Baku, p. 160 (published in Russian).
7. Measures of potential strengthening on the climate change in priority fields of Azerbaijan economy, 2nd phase. Edited by M.R. Mansimov, Baku, "Capp-Poligraf", 2001, p. 64 (published in Russian).

CHAPTER 21

ECOLOGICAL CONSIDERATIONS RELATED TO URANIUM EXPLORATION AND PRODUCTION

P.G. KAYUKOV
*JSC "Volkovgeology", 156a, Bogenbay batyr str.,
Almaty, 050012, Kazakhstan*

Abstract: This paper is considering modern ecological and radiation protection requirements in Kazakhstan, which is in need due to the Kazakhstan's uranium exploration, exploitation and production. Some specific problems appearing from *in situ* leaching production and geological exploration are presented, and proposed measures to overcome them are discussed.

Keywords: uranium exploration, uranium exploitation, uranium production, radiation protection requirements, Kazakhstan

1. Introduction

Ecological requirements to uranium exploration and production increase yearly both qualitatively and in number. During the last 20 years, many projects have been closed or stopped due to the new requirements. Industrial activities are not allowed to continue without ecological considerations. Kazakhstan government states that their aim is environmental protection favourable to human live and health. Therefore, there are set some normative standards of:

- Limited permissible residues of pollutants
- Waste disposals
- Detrimental physical impacts, and exemption of natural resources.

It is expected that the future results of following these standards, will be the steady development of Kazakhstan.

2. Principles

For the standards to be carried out, the following principles are put into acting laws:

- Obligatory performance of preventive measures against polluting the environment and environmental impacts in any form

- Inevitable responsibility for environmental delinquencies
- Obligatory compensation of impact to environment
- Using of the best ecologically pure technology taking care of natural resources and environment
- Any planned industrial activity is seen as dangerous for the environment and, therefore, environmental impact assessment is obligatory before making any decision about the fulfilment of the activity.

It is of importance that it is realized that these requirements have to be performed at all levels of industrial activities on a regional, national and international scale. So far, the world society doesn't reconcile these ecological problems in the same way they did after the Chernobyl accident. In fact, steady legislature has been created in Kazakhstan that gives direction to any industrial activity in such way that environment protection are provided, favourable to life and health of people. Ecological arrangements of any industrial activity start with a license. First of all, it is carried out in the industrial branches of atomic energy and precursors (toxic chemical elements and compounds). Then, uranium exploration and production can be performed after making a contract with the government of Kazakhstan. To get the contract, the industry needs to provide some ecological requirements, such as the creation of a special fund for financing environmental protection measures after closure of an enterprise. All predesign and design documentation should be passed through the national ecological expertise for approval.

During the realization of a project, a deposit contractor must justify an annual plan for the mining work, considering all requirements for environment protection. Here, it should be noticed that the ecological part of a design, environmental protection or environmental impact assessment, must be agreed through some public discussions. Yearly, each deposit and nature contractor must confirm their intentions to provide environment protection by appling for permission to pollute the environment. To get such permission, the contractor provides the regulatory body the following papers:

- Annual monitoring reports
- Calculations of volumes of air and water pollutants and waste
- Risk insurance
- Conclusions of ecological expertise on new designs
- Supplementary data about residues and waste.

Furthermore, the nature contractor needs to provide documentation on residues, waste and environmental monitoring. As uranium exploration and production will be increased more than three times the coming years, the negative impact on environment will also increase. So there is a need to follow the updated ecological requirements based on SRS-99 (Standards of Radiation Safety – 1999), SHRPRS-2003 (Sanitary Hygienic Requirements on Providing Radiation Safety – 2003) and SHR-ISL-99 (Sanitary Hygienic Rules-*In Situ* Leaching – 1999).

3. Uranium Production

Uranium production based on *in situ* leaching has been carried out at depths from 100 to 700 m in Paleogene and Cretaceous aquifers of different thickness. One specific property of the uranium ore – containing a series of radionuclides – is very important. Average specific radioactivity of the ores in these aquifers changes from 35 to 89 Bq/g which will exceed the discharge level for natural uranium. Therefore, both drilled ore cores and the mud are highly radioactive. Also the water is radioactive within the zones of oxidation in these aquifers (Table I). In accordance with i.185 of SHRPRS-2003 the level of radium, polonium and lead will exceed the exemption levels more than ten times, meaning that unused obtained water must be treated as radioactive waste.

All steps in the construction of technological wells for *in situ* leaching uranium production, especially during drilling ore intervals, enlarging well holes within radioactive intervals and testing the wells, may have a negative impact on the environment. Many specialists continue to consider that uranium production by *in situ* leaching stays dangerous for underground media, however that is a topic of another article. Therefore, construction and investigations of wells for uranium exploration and production lead to increased amounts of radioactive drilling mud and water-sand pulp.

TABLE I. Radioactivity in underground water within the zones of oxidation (Bq/l)

Radionuclide	Minimal activity	Maximal activity	Exemption levels, SRS-99
U	0.02	1.8	3.1
Ra-226	0.04	45.7	0.5
Pb-210	0.02	16.0	0.2
Po-210	0.02	1.8	0.12
Ra-228	0.04	0.9	0.37

It is common that multiple drilling solutions become radioactive when used and further use hampers the possibility to make gamma-logging in the wells. The radioactivity concentration of drilling mud can be 600 Bq/kg above the background levels in soil. The radioactivity concentration of water-sand pulp from wells within active production blocks increases three times compared to radioactivity concentration in water-sand pulp from blocks prepared for production. Besides sumps of wells, both drilling mud disposals and groundwater are polluted by the drilling waste (Table II). Runoff from drilling waste disposals increases the radionuclide concentration in the nearby soil, giving 600 Bq/kg above background levels and salinity up to 0.6% higher than background (Table II).

TABLE II. Radiometric and spectrometric analyses of samples obtained from drilling waste disposals, Bq/kg

Site	Sampling object	Gross-alpha	Th-234	Ra-226	Pb-210	U-235	Th-228	Ra-228
Kanzhugan	Soil	4,330	296	60	118	7.8	30	37
	Drilling mud	12,950	2,271	257	1,969	55.3	29	18
Karamurun	Soil	3,348	61	26	108	<6.8	10	26
	Drilling mud	4,281	114	71	176	7.4	16	18

TABLE III. Results from chemical analyses of samples obtained from drilling waste disposals

Site	Sampling object	pH	Salinity, %
Kanzhugan	Soil	8.97	0.078
	Drilling mud	8.17	0.946
Karamurun	Soil	8.28	1.561
	Drilling mud	7.70	1.907

Space Shuttle Survey of the acting *in situ* leaching polygons shows considerable watering production blocks, especially within artesian contours. Radioactive contamination of groundwater may be limited due to the sorption capacity of soils, especially as clay particles in soils may adsorb radioactive nuclides (from field investigations of run-off from artesian wells). By considering all possible environmental impact (Table III), the below environmental protection measures should be performed:

- Clearance and disposal of water-sand pulp into sand-precipitation basin ore additional wells
- Selective withdrawal of contaminated ground and drilling mud, its consolidation, removal and transfer into radioactive waste disposals
- Increasing radiometric control of the impact of drilling mud disposals on the environment

3. Conclusion

Radioactivity concentrations in groundwater and product waste from technological lines are above the radioactive standards set for uranium exploration and production in Kazakhstan. Therefore, there is a need to make good environmental protection measures. The proposed preventive measures discussed in this paper, will give future protection of the environment from uranium exploration and production in Kazakhstan.

CHAPTER 22

JOINT NORWEGIAN AND KAZAKH FIELDWORK IN KURDAY MINING SITE, KAZAKHSTAN, 2006

G. STRØMMAN[1], B.O. ROSSELAND[1],
J. ØVERGAARD[1], M. BURKITBAYEV[2],
I.A. SHISHKOV[3], B. SALBU[1]
[1]*Department of Plant and Environmental Sciences,*
Norwegian University of Life Sciences, P.O. Box 5003,
N-1432 AAS, Norway
[2]*Department of Inorganic Chemistry,*
Al-Farabi, Kazakh National University,
Kazhakstan Republic
[3]*Central Analytical Laboratory of Volkhovgeologiya,*
Kyrgyzstan

Abstract: A Joint Norwegian Kazakh and Kyrgy project has been established related to the contamination associated with uranium mining and tailing in the former USSR republics. The aims of the project are to assess long term consequences from radioactive and trace metal contamination, in particular associated with uranium mining and tailing, at selected sites in Kazakhstan and Kyrgyzstan, to evaluate the needs for alternative countermeasures, and to strengthen the scientific competence as well as the national authorities with respect to infrastructure and legislation for radiation protection. The present paper describes the first joint fieldwork, which took place in June 2006, at the uranium mining area at Kurday, Kazakhstan. Radon detectors were place indoor and outdoor. A series of samples were collected and in situ fractionation of radionuclides in lake, river and well water were collected. A broad analytical programme, including the analysis of radionuclides and metals, will be implemented in 2006/2007.

Keywords: uranium mining, radionuclides and metals, joint fieldwork, Kurday, joint Norwegian-, Kazakh-, Kyrgy-project

1. Introduction

The purpose of the Joint project between Norway, Kazakhstan and Kyrgyzstan, is to assess the long term consequences from radioactive and trace metal contamination at selected sites in Kazakhstan and Kyrgyzstan, sites which are in particular associated with uranium mining and tailing. The project period is three years, with the possibility of extension.

The project is a joint collaboration between responsible scientists and authorities in Norway, Kazakhstan and Kyrgyzstan. Participants from Norway include scientists from the University of Life Sciences in Norway (UMB) and the Norwegian Radiation Protection Authority (NRPA). Participants from Kazakhstan include scientists from Al-Farabi Kazakh National University (Al-Farabi University), Institute of Nuclear Physics of the National Nuclear Centre of the Republic of Kazakhstan (INP NNC KZ) and Central Scientific Research Analytical Laboratory, Volkhovgeologia. Participants from Kyrgyzstan include The Central Research Laboratory (CRL). The overall objecttives of the joint collaboration is to

- Assess long term consequences from radioactive and trace metal contamination, in particular associated with uranium mining and tailing, at selected sites in Kazakhstan and Kyrgyzstan
- Evaluate the needs for alternative countermeasures
- Strengthen the scientific competence as well as the national authorities with respect to infrastructure and legislation for radiation protection.

The main goals of the first joint fieldwork, which took place in Kurday, Kazakhstan June 2006, gamma dose measurements, placement of radon detectors, both in field and in the dwellings of Kurday as well as indoor in the nearby village as well as the collection of a series of soil, water, sediment, vegetation and fish samples.

2. Description of the Kurday Mining Site

Kurday is a small village situated in the south-east of Kazakhstan, not far from the border of Kyrgyzstan. The uranium mining site at Kurday opened in 1953, and was closed in the end of 1950s. During mining, the uranium containing rocks were crushed and transport from the site, and the site has been left without remediation until recently. Thus, the radionuclide and metal contamination associated with the mining activities within the site and a possible impact of the environment has never been investigated. The field investigated included Dead Lake, an artificial lake, (Fig. 1) created as a result of the mining activities. Dead Lake is surrounded by steep hills and a mountain plateau, which has been covered by crushed bedrock wastes. The plateau has recently been covered by about 1 m of clay, to prevent environmental pollution from the uranium rich crushed bedrock. However, the clay layers show already ruptures due to high altitude wind erosion. At the edges of the plateau, several landslides have also occurred revealing the presence of the underneath crushed bedrock wastes. Small creeks and springs, carrying groundwater were observed close to several of the landslide areas.

3. Sampling and Treatment of Samples

Samples were collected to quantify the heavy metal and radionuclide concentrations in water, sediments, soil, vegetation and in fish collected from the Dead Sea. Water

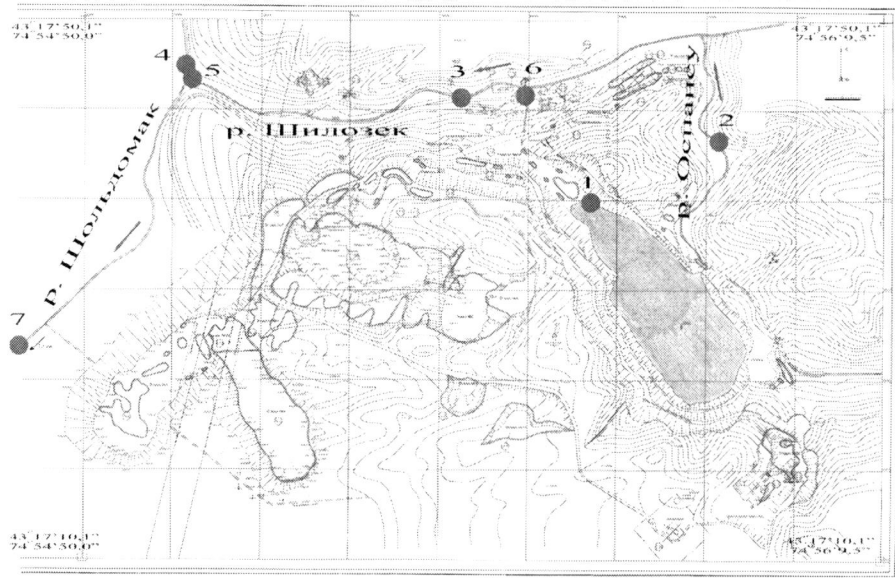

Figure 1. Map of the mining site of Kurday. Points 1–6 indicate the water sampling locations

Figure 2. Fish sampling by gillnet, by assistance from of a local fisherman

samples were collected in Dead Lake and in rivers and springs surrounding the mining area. As radionuclides and metals can be present in different physico-chemical forms, ranging from simple ions, to colloids and particles [1], the water samples were fractionated *on-site* at the different locations labelled 1–7 in Fig. 1. Fish samples were collected in Dead Lake with support of a local fisherman (Fig. 2). Soil and vegetation samples were collected around the Dead Lake, at the hills and at the plateau.

Outdoor, radon detectors were installed around Dead Lake, at the hills and in a grid pattern at the plateau. Radon detectors were also placed indoor in the dwellings of Kurday, and outdoor in the garden of the dwellings.

Gamma doses were measured with a proportional counter at the sampling sites, about 1 m above ground and immediately above ground. Gamma doses were also measured in all dwellings. Results from the different gamma dose meters were compared for calibration purposes. The sampling positions were registered with GPS and the positions were marked on the map (Fig. 3).

3.1. WATER AND SEDIMENTS

Water samples (1–5 L) were taken for total measurements at locations 1–7 (Fig. 1). Location 1 is the Dead Lake at the entry point, location 2 is a small creek draining from the U containing rock area and fed by the ground water, location 3 is the river draining the affected area, location 4 and 5 are at the confluence of 2 rivers draining the area, location 5 is a ground water spring and location 7 is river representing the overall output from the area. In addition to the collection of total samples, *at site* fractionation techniques were applied to distinguish between radionuclides and metals present as low molecular mass (LMM) species, or associated with colloids or particles. Combined size and charge fractionation of radionuclides and metals in water was performed using of hollow fibre (Amicon, nominal molecular size cut-off of 10 kDa) interfaced with ion chromatography (cation resin Chelex) [1]. All the fractions are currently analyzed with respect to radionuclides) and heavy metal concentrations and isotopic composition (uranium-isotopes). The concentration of radium (5 L) will also be determined.

Sediments were collected using a polyethylene container at the water sample locations. In lab, sediments were air dried, homogenised and grounded prior to dissolution and measurements.

3.2. FISH

Surprisingly, many fish were caught by using gillnets, and three fish species were present. For fish still alive, they were killed with a blow to the head and blood samples were collected immediately *on-site* at Dead Lake (Fig. 4). The fish were dissected and the organs (liver, kidney, fillet, gills) were collected and preserved (Fig. 5). The heavy metal concentration in liver, gills and kidney will be determined. In addition, filet samples were collected for the delta N-15 and delta C-14 to determine the trophic levels of the fish and to obtain information about the food sources. Furthermore, the $^{234}U/^{238}U$ isotope ratio will be measured at the Al-Farabi University and in Norway to identify the

Figure 3. Satellite map of the Kurday mining site. Blue squares indicates houses, yellow circles indicate sampling positions and red dots indicate positions of the radon detectors

source of uranium in fish. Environmental contaminants (POP's) will also be determined in fish filets.

3.3 SOIL

Soil samples were collected at different locations around Dead Lake, at the plateau and surrounding areas. Activity of radionuclides and the heavy metal concentration will be measured I sliced samples. Sequential extraction of soils will be carried out to identify whether radionuclides and metals are reversibly associated to soil components.

3.4 VEGETATION

Different plant materials were collected and the plant species will be identified at the Al-Farabi University. In lab, the plant samples were dried at room temperature, grinded homogenised and ashed prior to the determination of radionuclides and metals.

Figure 4. On-site blood-sampling and dissection of fish at the Dead Lake

Figure 5. Dissection of fish continued in the evening at the campus

3.5. RADON DETECTORS

Radon detectors were placed at different locations in the field. A the plateau, the Rn detectors were placed in a rectangular grid system. Radon detectors were also placed inside and in the gardens of about 20 dwellings in Kurday (Fig. 6). Nearly all detectors were collected two months later, transported to Norway and analyzed at NRPA, Norway.

3.6. PARTICLES

Using gamma detectors, hot spots and individual stones containing uranium and daughters could easily be identified in the field. In the lab, heterogeneities were easily identified by autoradiography. Using electron microscopy, the spatial distribution of individual daughters within hot spots will most probably be obtained. Sequential extraction will also be carried out to investigate if associated radionuclides could migrate from the mineral phases. Weathering experiments will be performed to estimate weathering rates and the remobilation potential of U and daughters [1].

Heavy metals in water, soil, fish and vegetation will be determined using Inductive Coupled Plasma mass spectrometry (ICP-MS), neutron activation analysis (NAA) and

Flow Injection Mercury System (FIMS). Radionuclides will be determined using Liquid Scintillation Counting (LSC), gamma spectrometry (GE-detector), ICP-MS and NAA. Isotope ratios of stable isotopes (delta N-15 and delta C-13) will be determined using Isotope Ratio Mass Spectrometry (IR-MS). Physical-chemical variables, pH, ionic strength, Total Organic Carbon (TOC), major cationic and anionic concentrations were measured in water samples.

Figure 6. Local food-store in Kurday where radon detectors were placed

4. Conclusion

A total of 5–10 L of water samples were collected at each sites in Fig. 1. Totally, 13 soil samples, 5 sediment samples and 11 vegetation samples were collected at different sites, and 40 fish samples were collected in Dead Lake. All the samples were sent to laboratories in Norway, at UMB and NRPA, and to Kazhakstan, at Al-Farabi University. When results are available, people living in the Kurday village as well as the authorities will be informed.

References

1. Salbu, B. (2000) Speciation of Radionuclides. *Encyclopaedia Analytical Chemistry.* Wiley, Chichester, 12993–13016.

LIST OF CONTRIBUTORS

AGEYEVA, T.I.
Institute of Nuclear Physics, Almaty, Kazakhstan Republic

AITMATOVA, D.I.
Institute of Physics and Mechanic of Rock Stones of National Academy Sciences, Bishkek, Kyrgyzstan Republic

ALEKHINA, V.M.
Institute of Physics NASc KR 265a, Chui prospect, Bishkek, 720071, Kyrgyz Republic

ALEKSANYAN, G.M.
Yerevan State University, Department of Geology, 1 Alex Manoogian str., 0025, Yerevan, Armenia

ARAKELYAN, V.
The Yerevan Institute of Physics, Yerevan, Armenia

ARTEMOV, S.V.
Institute of Nuclear Physics AS RU

ATOYAN, V.
Armenian Nuclear Power Plant, Medzamor, Republic of Armenia

BAKHTIAR, S.N.
Institute of Dynamic Change, 2135 Ascot Dr. #28, Moraga, CA 94556

BARBER, D.S.
Geosciences and Environment Center Sandia National Laboratories, Cooperative, Monitoring Center, USA

BEZZUBOV, N.I.
SE "Vostokredmet", Kalinina Str., Sogd Oblast, Khudjand City, Tajikistan Republic

BURKITBAYEV, M.
Department of Inorganic Chemistry, Al-Farabi Kazakh National University, Almaty, Kazakhstan Republic

CHERNYKH, E.E.
Institute of Nuclear Physics NNC RK, Ibragimov str. 1, 050032, Almaty, Kazakhstan Republic

CHKHARTISHVILI, M.S.
Scientific Center for Radiobiology and Radiation Ecology, Georgian Academy of Sciences, 51 Telavi Str., 0103, Tbilisi, Georgia

DANILOVA, E.A.
Institute of Nuclear Physics AS RU

DAVLATSHOEV, T.
Physical-Technical Institute of the Academy of Sciences of Republic of Tajikistan (PhTI)

DJURAEV, A.
Atomic Energy Agency, Tajik Academy of Sciences, Dushanbe, Tajikistan

DUBASOV, YU.V.
RPA "V.G. Khlopin Radium Institute", 28, 2-nd Murinskii av., St-Petersburg, 194021, Russia

DZURAEV, AN.A.
Physical-Technical Institute of the Academy of Sciences of Republic of Tajikistan (PhTI)

HOWLETT, J.G.
UCD School of Physics, University College Dublin, Belfield, Dublin 4, Ireland

IDRISOVA, S.
Institute of Physics NASc KR 265a, Chui prospect, Bishkek, 720071, Kyrgyz Republic

IVANENKO, D.A.
Institute of Physics NASc KR 265a, Chui prospect, Bishkek, 720071, Kyrgyz Republic

JURAEV, A.A.
Physical-Technical Institute of the Academy of Sciences of Republic of Tajikistan (PhTI).

KADYRZHANOV, K.K.
Institute of Nuclear Physics NNC RK, Ibragimov str. 1, 050032, Almaty, Kazakhstan Republic

KAYUKOV, P.G.
JSC "Volkovgeology", 156a, Bogenbay batyr str., Almaty, 050012, Kazakhstan

KAZAKOV, S.V.
Nuclear Safety Institute of the Russian Academy of Sciences (IBRAE RAS), 52 b. Tulskaya str., 115191, Moscow, Russia

KHAZEKHBER, S.
Institute of Nuclear Physics NNC RK, Ibragimov str. 1, 050032, Almaty, Kazakhstan Republic

KIST, A.A.
Institute of Nuclear Physics AS RU

LIST OF CONTRIBUTORS

KUYANOVA, Y.
Department of Inorganic Chemistry, Al-Farabi Kazakh National University, Almaty, Kazahstan Republic

LEÓN VINTRÓ, L.
UCD School of Physics, University College Dublin, Belfield, Dublin 4, Ireland

LESPUKH, E.E.
Institute of Nuclear Physics AS RU

LUKASHENKO, S.N.
Institute of Radiation Safety and Ecology, National Nuclear Centre, Kurchatov, Kazakhstan Reublic

MAGERRAMOV, G.
Geology Institute of Azerbaijan National Academy of Sciences, 29a H.Javid av., AZ1143, Baku, Azerbaijan

MAMATIBRAIMOV, S.
Institute of Physics NASc KR 265a, Chui prospect, Bishkek, 720071, Kyrgyz Republic

MAMEDOV, R.F.
Geology Institute of Azerbaijan National Academy of Sciences, 29a H. Javid av., AZ1143, Baku, Azerbaijan

MITCHELL, P.I.
UCD School of Physics, University College Dublin, Belfield, Dublin 4, Ireland

ØVERGAARD, J.
Dept of Plant and Environmental Sciences, orwegian University of Life Sciences, P.O. Box 5003, N-1432 AAS, Norway

PASSELL, H.D.
Sandia National Laboratories, P.O. Box 5800, Albuquerque, NM87185-1373, USA

PETROV, V.A.
Technical University, 492000, Ust-Kamenogorsk, Kazakhstan Republic

PODENEZHKO, V.V.
Institute of Nuclear Physics, Almaty, Kazakhstan Republic

POHL, P.
Geoecology Department, University of Potsdam, Potsdam, Germany

POZNYAK, V.L.
Institute of Nuclear Physics NNC RK, Ibragimov str. 1, 050032, Almaty, Kazakhstan Republic

PRIEST, N.D.
School of Health and Social Sciences, Middlesex University, Queensway,
Enfield EN3 4SA, UK

PYUSKYULYAN, K.
Armenian Nuclear Power Plant, Medzamor, Republic of Armenia

RADYUK, G.A.
Institute of Nuclear Physics AS RU

RADYUK, R.I.
Institute of Nuclear Physics AS RU

RAZIKOV, Z.A.
SE "Vostokredmet", Kalinina Str., Sogd Oblast,
Khudjand City, Tajikistan Republic

RISOLUTI, P.
International Atomic Energy Agency, Wagramer Strasse 5, P.O. Box 100,
A-1400 Vienna Austria

ROSSELAND, B.O.
Department of Plant and Environmental Sciences, Norwegian University
of Life Sciences, P.O. Box 5003, N-1432 AAS, Norway

SAGHATELYAN, A.
Institute of Noosphere Researches of AS of Armenia

SALBU, B.
Department of Plant and Environmental Sciences, Norwegian University
of Life Sciences, P.O. Box 5003, N-1432 AAS, Norway

SALIKHBAEV, U.S.
Institute of Nuclear Physics AS RU

SHAMAEVA, A.A.
Nuclear Safety Institute of the Russian Academy of Sciences (IBRAE
RAS), 52 b. Tulskaya str., 115191, Moscow, Russia

SHISHKOV, I.A.
Central Analytical Laboratory of Volkhovgeologiya, Kyrgyzstan

SOLODUKHIN, V.P.
Institute of Nuclear Physics NNC RK, Ibragimov str. 1,
050032, Almaty, Kazakhstan Republic

STARODUMOV, O.I.
Institute of Physics NASc KR 265a, Chui prospect, Bishkek, 720071,
Kyrgyz Republic

STEPANETS, O.V.
Vernadsky Institute of Geochemistry and Analytical Chemistry of RAS, 19
Kosygin str., 11999, Moscow, Russia

LIST OF CONTRIBUTORS

STRILCHUK, YU.G.
Institute of Radiation Safety and Ecology, National Nuclear Centre, Kurchatov, Kazakhstan Republic

STRØMMAN, G.
Department of Plant and Environmental Sciences, Norwegian University of Life Sciences, P.O. Box 5003, N-1432 AAS, Norway

SUOZZI, D.
Geosciences and Environment Center Sandia National Laboratories, Cooperative, Monitoring Center, USA

TILLOBOEV, KH.I.
SE "Vostokredmet", Kalinina Str., Sogd Oblast, Khudjand City, Tajikistan Republic

TOLSTIKHIN, G.M.
Institute of Physics NASc KR 265a, Chui prospect, Bishkek, 720071, Kyrgyz Republic

TULEUSHEV, A.ZH.
Institute of Nuclear Physics, Almaty, Kazakhstan Republic

VALYAEV, A.N.
Nuclear Safety Institute of the Russian Academy of Sciences (IBRAE RAS), 52 b. Tulskaya str., 115191, Moscow, Russia

VASILEVA, V.S.
Institute of Nuclear Physics AS RU

VASILIEV, I.A.
Institute of Physics NASc KR 265a, Chui prospect, Bishkek, 720071, Kyrgyz Republic

WAGGITT, P.W.
Waste Safety Section, International Atomic Energy Agency, Wagramerstrasse 5, A-1400, Vienna, Austria

YULDASHEV, B.S.
Institute of Nuclear Physics AS RU

YUNUSOV, M.M.
SE "Vostokredmet", Kalinina Str., Sogd Oblast, Khudjand City, Tajikistan Republic

ZAPAROV, E.A.
Institute of Nuclear Physics AS RU

ZURAVLEV, A.A.
Institute of Nuclear Physics AS RU

Printed in the United States
112963LV00001B/19-27/P